Autodesk® Inventor® 2012 School Edition

Autodesk® Inventor® 2012 School Edition

DANIEL T. BANACH
TRAVIS JONES
ALAN J. KALAMEJA

DELMAR
CENGAGE Learning™

Australia • Brazil • Japan • Korea • Mexico • Singapore • Spain • United Kingdom • United States

Autodesk® Inventor® 2012 School Edition

Daniel T. Banach, Travis Jones, and Alan J. Kalameja

Vice President, Editorial: Dave Garza

Director of Learning Solutions: Sandy Clark

Acquisitions Editor: Stacy Masucci

Managing Editor: Larry Main

Senior Product Manager: Mary Clyne

Editorial Assistant: Andrea Timpano

Vice President, Marketing: Jennifer Baker

Marketing Director: Deborah Yarnell

Associate Marketing Manager: Jillian Borden

Production Director: Wendy Troeger

Senior Content Project Manager: Angela Sheehan

Art Director: David Arsenault

Technology Project Manager: Joe Pliss

© 2012 Delmar, Cengage Learning

ALL RIGHTS RESERVED. No part of this work covered by the copyright herein may be reproduced, transmitted, stored, or used in any form or by any means graphic, electronic, or mechanical, including but not limited to photocopying, recording, scanning, digitizing, taping, Web distribution, information networks, or information storage and retrieval systems, except as permitted under Section 107 or 108 of the 1976 United States Copyright Act, without the prior written permission of the publisher.

Autodesk, AutoCAD, Inventor, and the Autodesk logo are registered trademarks of Autodesk. Delmar Cengage Learning uses "Autodesk Press" with permission from Autodesk for certain purposes.

For product information and technology assistance, contact us at
Cengage Learning Customer & Sales Support, 1-800-354-9706
For permission to use material from this text or product,
submit all requests online at **www.cengage.com/permissions**
Further permissions questions can be e-mailed to
permissionrequest@cengage.com

Library of Congress Control Number: 2011928766

ISBN-13: 978-1-111-64847-3

ISBN-10: 1-111-64847-6

Delmar
5 Maxwell Drive
Clifton Park, NY 12065-2919
USA

Cengage Learning is a leading provider of customized learning solutions with office locations around the globe, including Singapore, the United Kingdom, Australia, Mexico, Brazil, and Japan. Locate your local office at: **international.cengage.com/region**

Cengage Learning products are represented in Canada by Nelson Education, Ltd.

To learn more about Delmar, visit **www.cengage.com/delmar**

Purchase any of our products at your local college store or at our preferred online store **www.cengagebrain.com**

Notice to the Reader
Publisher does not warrant or guarantee any of the products described herein or perform any independent analysis in connection with any of the product information contained herein. Publisher does not assume, and expressly disclaims, any obligation to obtain and include information other than that provided to it by the manufacturer. The reader is expressly warned to consider and adopt all safety precautions that might be indicated by the activities described herein and to avoid all potential hazards. By following the instructions contained herein, the reader willingly assumes all risks in connection with such instructions. The publisher makes no representations or warranties of any kind, including but not limited to, the warranties of fitness for particular purpose or merchantability, nor are any such representations implied with respect to the material set forth herein, and the publisher takes no responsibility with respect to such material. The publisher shall not be liable for any special, consequential, or exemplary damages resulting, in whole or part, from the readers' use of, or reliance upon, this material.

Printed in the United States of America
1 2 3 4 5 6 7 15 14 13 12 11

CONTENTS

Introduction ix

CHAPTER 1 GETTING STARTED 1

Introduction 1 • Objectives 1 • Getting Started with Autodesk Inventor 2 • User Interface 2 • Inventor Application Menu 4 • Ribbon 5 • Quick Access Toolbar 6 • File Information 9 • Application Options 13 • Exercise 1-1: User Interface 14 • Repeat Last Command 19 • Help System 20 • Projects in Autodesk Inventor 21 • Exercise 1-2: Projects 27 • Viewpoint Options 29 • Exercise 1-3: Viewing a Model 37 • Checking Your Skills 42

CHAPTER 2 SKETCHING, CONSTRAINING, AND DIMENSIONING 43

Introduction 43 • Objectives 43 • Part and Sketch Application Options 43 • Units 46 • Templates 46 • Creating a Part 47 • Step 1—Sketch the Outline of the Part 50 • Exercise 2-1: Creating a Sketch With Lines 62 • Exercise 2-2: Creating a Sketch With Tangencies 64 • Step 2—Constraining the Sketch 65 • Construction Geometry 69 • Exercise 2-3: Adding and Displaying Constraints 72 • Step 3—Adding Dimensions Manually 76 • Exercise 2-4: Dimensioning a Sketch 83 • Applying Your Skills 85 • Checking Your Skills 86

CHAPTER 3 CREATING AND EDITING SKETCHED FEATURES 87

Introduction 87 • Objectives 87 • Understanding Features 87 • Understanding the Browser 89 • Switching Environments 90 • Model Commands 91 • Direct Manipulation 91 • Extrude a Sketch 93 • Exercise 3-1: Extruding a Sketch 99 • Linear Diameter Dimensions 100 • Revolve a Sketch 102 • Exercise 3-2: Revolving a Sketch 107 • Editing a Feature 109 • Exercise 3-3: Editing Features and Sketches 113 • Sketched Features 114 • Define the Active Sketch Plane 115 • Slice Graphics 117 • Exercise 3-4: Sketch Features 117 • Projecting Part Edges 119 • Project Edges 120 • Exercise 3-5: Project Geometry 122 • Part Material, Properties, and Color 124 • Applying Your Skills 125 • Checking Your Skills 128

CHAPTER 4 CREATING PLACED FEATURES 129

Introduction 129 • Objectives 129 • Fillets 130 • Chamfers 137 • Exercise 4-1: Creating Fillets and Chamfers 140 • Holes 143 • Exercise 4-2: Creating Holes 148 • Shelling 153 • Exercise 4-3: Shelling a Part 155 • Work Features 157 • Exercise 4-4: Creating Work Axes 159 • Creating Work Planes 162 • Exercise 4-5: Creating Work Planes and a UCS 169 • Patterns 175 • Exercise 4-6: Creating Rectangular Patterns 181 • Exercise 4-7: Creating Circular Patterns 183 • Exercise 4-8: Creating a Pattern Along a Nonlinear Path 184 • Applying Your Skills 186 • Checking Your Skills 187

CHAPTER 5 CREATING AND EDITING DRAWING VIEWS 188

Introduction 188 • Objectives 188 • Drawing Sheet Preparation 191 • Save Drawing Data to a Template 193 • Creating Drawing Views 193 • Creating a Base View 194 • Exercise 5-1: Creating a Multiview Drawing 199 • Exercise 5-2: Creating Auxiliary, Section, and Detail Views 213 • Exercise 5-3: Creating a Break View 218 • Editing Drawing Views 219 • Exercise 5-4: Editing Drawing Views 222 • Annotations 224 • Drawing Standard and Styles 225 • Exercise 5-5: Adding Centerlines 231 • Adding Dimensions to a View 234 • Exercise 5-6: Adding Dimensions 241 • Creating Baseline Dimensions 246 • Creating Chain Dimensions 249 • Exercise 5-7: Creating Baseline and Chain Dimensions 251 • Exercise 5-8: Adding Annotations 257 • Opening a Model from a Drawing 260 • Opening a Drawing from a Model 261 • Applying Your Skills 261 • Checking Your Skills 262

CHAPTER 6 CREATING AND DOCUMENTING ASSEMBLIES 264

Introduction 264 • Objectives 264 • Creating Assemblies 265 • Creating Parts in Place 267 • Editing a Component in Place 270 • Opening and Editing Assembly Components 272 • Degrees of Freedom (DOF) 272 • Assembly Constraints 274 • ALT + Drag Constraining 284 • Editing Assembly Constraints 286 • Exercise 6-1: Assembling Parts 288 • Additional Constraint Commands 295 • Assembly Browser Commands 299 • Adaptivity 301 • Assembly Sections 303 • Exercise 6-2: Designing Parts in the Assembly Context 304 • Patterning Components 308 • Exercise 6-3: Patterning Components 312 • Analysis Commands 315 • Exercise 6-4: Analyzing an Assembly 317 • Driving Constraints 319 • Exercise 6-5: Driving Constraints 322 • Creating Presentation Files 324 • Autodesk Inventor Publisher 330 • Exercise 6-6: Creating Presentation Views 332 • Creating Drawing Views from Assemblies and Presentation Files 336 • Creating Balloons 336 • Parts Lists 343 • Parts List Commands 345 • Exercise 6-7: Creating Assembly Drawings 348 • Applying Your Skills 354 • Checking Your Skills 359

CHAPTER 7 ADVANCED MODELING TECHNIQUES 360

Introduction 360 • Objectives 360 • Adjustable Sections Views in a Part File 361 • Design View Representation in a Part File 362 • Embossed Text and Closed Profiles 363 • Exercise 7-1: Creating Text and Emboss Features 367 • Dimension Display, Relationships, and Equations 372 • Parameters 375 • Exercise 7-2: Relationships and Parameters 381 • Sweep Features 388 • Exercise 7-3: Creating Sweep Features 392 • Project to 3D Sketch 398 • Exercise 7-4: 3D Sketch—Sweep Features 398 • Loft Features 401 • Exercise 7-5: Creating Loft Features 407 • Mirror Features 411 • Suppressing Features 412 • Reordering a Feature 413 • Feature Rollback 413 • Content Center 415 • Design Accelerator 416 • Stress Analysis 417 • Exercise 7-6: Running a Stress Analysis on a Part 426 • Exercise 7-7: Running a Stress Analysis on an Assembly 428 • Applying Your Skills 429 • Checking Your Skills 431

Index 433

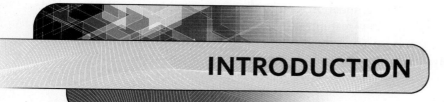

INTRODUCTION

INTRODUCTION

Welcome to the *Autodesk® Inventor® 2012 School Edition* manual. This manual provides a thorough coverage of the features and functionalities offered in Autodesk Inventor.

Each chapter in this manual is organized with the following elements:

Objectives. Describes the content and learning objectives.

Topic Coverage. Presents a concise, thorough review of the topic.

Exercises. Presents the workflow for a specific command or process through illustrated, step-by-step instructions.

Checking Your Skills. Tests your understanding of the material using True/False and short answer.

NOTE TO THE LEARNER

Autodesk Inventor is designed for easy learning. Autodesk Inventor's help system provides you with ongoing support as well as access to online documentation.

As described above, each chapter in this manual has the same instructional design, making it easy to follow and understand. Each exercise is task-oriented and based on real-world mechanical engineering examples.

User Prerequisites

It is recommended that you have a working knowledge of Microsoft® Windows XP Professional®, Windows Vista™, or Windows 7™.

Manual Objectives

The primary objective of this manual is to provide instruction on how to create part and assembly models, document those designs with drawing views, and automate the design process.

Upon completion of all chapters in this manual, you will be proficient in the following tasks:

- Basic and advanced part modeling techniques
- Drawing view creation techniques
- Assembly modeling techniques

While working through these materials, we encourage you to make use of the Autodesk Inventor help system, where you may find solutions to additional design problems that are not addressed specifically in this manual.

Manual Description

This manual provides the foundation for a hands-on course that covers basic and advanced Autodesk Inventor features used to create, edit, document, and parts and assemblies. You learn about the part and assembly modeling tools through print documentation and through the real-world exercises in this manual.

AUTODESK® INVENTOR® 2012 SCHOOL EDITION AND PROJECT LEAD THE WAY, INC.

This text resulted from a partnership forged between Delmar Cengage Learning and Project Lead The Way, Inc. in February 2006. As a not-for-profit foundation that develops curriculum for engineering, Project Lead The Way, Inc. provides students with the rigorous, relevant, reality-based knowledge they need to pursue engineering or engineering technology programs in college.

The Project Lead The Way® curriculum developers strive to make math and science relevant for students by building hands-on, real-world projects in each course. PLTW® students will acquire extensive experience using Autodesk Inventor to develop, model, and test their designs and hands-on projects. With assistance from PLTW® Master Teachers and Curriculum Directors, this text has been developed to meet the specific needs of PLTW® students and instructors in Engineering Design, Principals of Engineering, Computer Integrated Manufacturing, Civil Engineering and Architecture, Aerospace Engineering, and Engineering Design and Development.

To support Project Lead The Way's® curriculum goals, and to support all teachers who want to develop project/problem-based programs in engineering and engineering technology, Delmar Cengage Learning is developing a complete series of texts to complement all of Project Lead the Way's® nine courses:

Gateway to Technology
Introduction to Engineering Design
Principles of Engineering
Digital Electronics
Aerospace Engineering
Biotechnical Engineering
Civil Engineering and Architecture
Computer Integrated Manufacturing
Engineering Design and Development

To learn more about Project Lead The Way's® ongoing initiatives in middle school and high school, please visit www.pltw.org.

In addition, the author and publisher wish to acknowledge the valuable contributions of Project Lead the Way® Master Teachers Robert Tuttle and Todd Dischinger. In addition to writing and illustrating the Stress Analysis section and exercises, they helped shape the content in this custom edition by reviewing the proposal, outline, and edited manuscript. We also thank Project Lead the Way's® Curriculum Director, Sam Cox, for reviewing the proposal and providing guidance on content selection.

EXERCISE FILES

The exercise files for each chapter can be downloaded from http://www.cengagebrain.com. Copy the folder Inv 2012 School folder and its subfolders to your hard drive.

Accessing a Student Companion Site from CengageBrain:

1. Go to: http://www.cengagebrain.com.
2. ENTER Banach, Autodesk Inventor 2012 School Edition or 1111648476 in the **Find your Textbook or Materials** search box, click the **Find** button.
3. At the Product page, click the **Access Now** button.
4. At the Companion Product page, click the **Student Resources** link.

Projects

Most people work on several projects at a time, with each project consisting of a number of files. To accommodate this, Autodesk Inventor uses projects to help organize related files and maintain links between files.

Each project has a *project file* that stores the paths to all files related to the project. When you attempt to open a file, Autodesk Inventor uses the paths in the current project file to locate other necessary files.

For convenience, a project file is provided with the *exercises*.

Using the Project File

Before starting an exercise, you must have completed the following steps:

1. Start Autodesk Inventor.
2. On the Get Started tab > Launch panel, click Projects.
3. In the Projects window, select Browse. Navigate to the folder where you placed the Exercises, and double-click the file *Inv 2012 School.ipj*.
4. The "Inv 2012 School" project will become the current project.
5. You can now start doing the exercises.

Projects are reviewed in more detail in Chapter 1.

NOTE

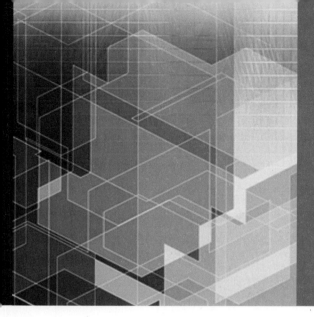

CHAPTER 1

Getting Started

INTRODUCTION

This chapter provides a look at the user interface, application options, starting commands, creating projects, and how to view models in Autodesk Inventor.

OBJECTIVES

In this chapter, you will gain an understanding of the following:

- The user interface
- How to open files
- How to create new files
- Different file types used in Autodesk Inventor
- Save options
- Application options
- How to issue commands
- The Help system
- Reasons for which a project file is used
- How to create a project file for a single user
- Different viewing and appearance commands

GETTING STARTED WITH AUTODESK INVENTOR

The default Autodesk Inventor screen looks similar to the following image. From here you can open existing files or create new files. The default screen can be changed to display the Open dialog box or the New dialog box, or to start a specified file, which can be set via the Application Options in the General tab under Start-up action.

FIGURE 1.1

USER INTERFACE

The default sketch environment of a part (*.ipt*) in the Autodesk Inventor application window is shown in the following image.

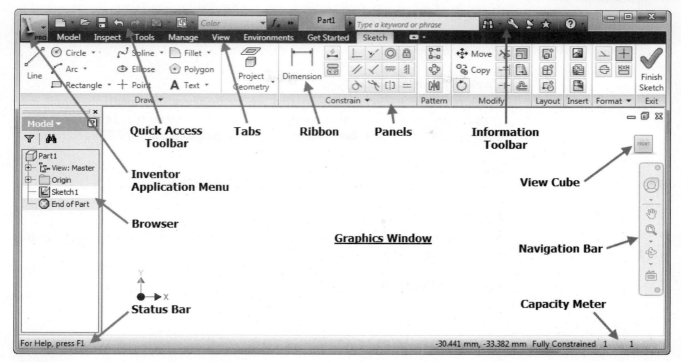

FIGURE 1.2

The screen is divided into the following areas:

Inventor Application Menu
Contains common commands for working with files.

Quick Access Toolbar
Accesses common commands as well as commands that can be added or removed.

Tabs
Changes available commands by clicking on a tab.

Ribbon
Accesses basic Windows and Autodesk Inventor commands. The set of commands in the ribbon changes to reflect the environment in which you are working.

Panels
Changes to show available commands for the current tab, click on a tab to display a set of new panels and commands.

Information Toolbar
Displays common help commands as well as Subscription services.

ViewCube
Displays current viewpoint and allows you to change the orientation of the view.

Navigation Bar
Displays common viewing commands. Viewing commands can be added by clicking the bottom drop arrow.

Capacity Meter
Displays how many occurrences (parts) are in the active document, the number of open documents in the current session, and how much memory is being used. Note: The capacity meter that shows memory usage is available only on 32-bit computers.

Status Bar
Views text messages about the current process.

Browser
Shows the history of how the contents in the file were created. The browser can also be used to edit features and components.

Graphics Window
Displays the graphics of the current file.

INVENTOR APPLICATION MENU
Besides selecting commands for working with files you can control how the recent or open documents are listed in the menu. The following image shows the functionality available from the Inventor Application Menu.

FIGURE 1.3

Recent Documents
Displays documents that were previously opened.

Open Documents
Displays documents that are opened.

Display
Controls what is displayed in the list: icons or images and what size.

Order to List Files
Controls the order that the files are listed: by Ordered List, by Access Date, by Size, by Type opened.

Pin to Menu
Click the push pin to keep the file in the list no matter when it was last opened.

Preview Image
Hovers the cursor over a file to see a larger image of the file when it was last saved.

> **NOTE**
> Double-click on the Inventor Application button, and a dialog box will appear asking you to save each unsaved document and then Inventor will close.

RIBBON

The ribbon displays commands that are relevant to the selected tab. The current tab is highlighted in green and is also green if it supports the current environment. The commands are arranged by panels. The most common commands are larger in size while less used commands are smaller and positioned to the right of the larger command. Tools may also be available in the drop list of a command or in the name of the panel. For example there is a drop list available under the Circle icon and the Draw panel as shown in the following image that shows the Sketch tab active in a part file.

FIGURE 1.4

The ribbon can be modified by right-clicking on the Panel; the following images show the available options.

FIGURE 1.5

- Ribbon Appearance – Changes how the Ribbon looks, turns text off, and changes the size of the commands
- Panels – Adds and removes Panels
- Customize User Commands – Adds commands to a User Panel
- Undock Ribbon – Allows the ribbon to be freely moved
- Docking Position – Changes the location of the Ribbon

QUICK ACCESS TOOLBAR

Tools can be removed or added from the Quick Access Toolbar.

To add a command follow these steps

1. Move the cursor over a command to add and right-click.
2. Click Add to Quick Access Toolbar as shown in the following image on the left.

To remove a command follow these steps

3. Move the cursor over the command to remove and right-click.
4. Click Remove from Quick Access Toolbar as shown in the following image on the right.

FIGURE 1.6

To gain more space for the Quick Access Toolbar it can be moved below the ribbon. Move the cursor over the Quick Access Toolbar and click Show Quick Access Toolbar below the Ribbon as shown in the following image on the left. Once the Quick Access Toolbar is placed below the ribbon, it can be moved back to its original location by moving the cursor over the Quick Access Toolbar and click Show Quick Access Toolbar above the Ribbon as shown in the following image on the right.

FIGURE 1.7

Open

To open files follow one of these techniques:

- Click the Open command in the Get Started tab in the Launch Panel as shown in the following image on the left.
- Click the Open command in the Quick Access Toolbar as shown in the following image on the left.
- Click the Inventor Application menu in the top-left corner and click Open as shown in the following image on the right.
- Press CTRL + O.

FIGURE 1.8

The Open dialog box will appear as shown in the following image. The directory that opens by default is set in the current project file. You can open files from other directories that are not defined in the current project file, but this is not recommended. Part, drawing, and assembly relationships may not be resolved when you reopen an assembly that contains components outside the locations defined in the current project file. Projects are covered later in this chapter.

FIGURE 1.9

New Files

Like the Open command there are many ways to create a new file.

To create a new Inventor file, follow one of these techniques:

- Click the New command in the Get Started tab in the Launch Panel as shown in the following image on the left.
- Click the New command in the Quick Access Toolbar as shown in the following image on the left.
- Click the Inventor Application button in the top-left corner and click New as shown in the following image in the middle.
- Press CTRL + N.
- To start a new file based on one of the default templates click the down arrow next to the New icon in the Quick Access Toolbar as shown in the following image on the right.

FIGURE 1.10

The New dialog box will appear as shown in the following image. Begin by selecting the type of file to create or one of the drafting standards named on the tabs, and then select a template for a new part, assembly, presentation file, sheet metal part, or drawing. If Autodesk Inventor Professional is installed, a Mold Design tab will exist.

FIGURE 1.11

Quick Launch

While working you may accidentally click the New command when you wanted to click the Open command or vice versa. In the New and Open dialog boxes, there is a Quick Launch area in the lower left corner of the dialog box.

FIGURE 1.12

FILE INFORMATION

While creating parts, assemblies, presentation files, and drawing views, data is stored in separate files with different file extensions. This section describes the different file types and the options for creating them.

File Types

The following section describes the main file types that you can create in Autodesk Inventor, their file extensions, and descriptions of their uses.

Part (.ipt)

Part files contain only one part, which can be either 2D or 3D.

Assembly (.iam)

Assembly files can consist of a single part, multiple parts, or subassemblies. The parts themselves are saved to their own part file and are referenced (linked) in the assembly file. See Chapter 6 for more information about assemblies.

Presentation (.ipn)

Presentation files show parts of an assembly exploded in different states. A presentation file is associated with an assembly, and any changes made to the assembly will be updated in the presentation file. A presentation file can be animated, showing how parts are assembled or disassembled. The presentation file extension is *.ipn*, but you save animations as AVI files. See Chapter 6 for more information about presentation files.

Sheet Metal (.ipt)

Sheet metal files are part files that have the sheet metal environment loaded. In the sheet metal environment, you can create sheet metal parts and flat patterns. You can create a sheet metal part while in a regular part. This requires that you load the sheet metal environment manually.

Drawing (.dwg and .idw)

Drawing files can contain 2D projected drawing views of parts, assemblies, and/or presentation files. You can add dimensions and annotations to drawing views. The parts and assemblies in drawing files are linked, like the parts and assemblies in assembly and presentation files. See Chapter 5 for more information about drawing views.

Project (.ipj)

Project files are structured XML files that contain search paths to locations of all the files in the project. The search paths are used to find the files in a project.

Opening Multiple Documents

You can open multiple Autodesk Inventor files at the same time by holding down the CTRL key and selecting the files to open as shown in the following image. Each file will be opened in its own window in a single Autodesk Inventor session. To switch between the open documents, click the file on the Windows menu. The files can also be arranged to fit the screen or to appear cascaded. If the files are arranged or cascaded, click a file to activate it. Only one file can be active at a time.

FIGURE 1.13

Document Tabs

When multiple documents are open in Inventor, each document appears in a tab in the lower left corner of the graphics window. The current document is represented with an "x" to the right of the file name and the tab's background is white. You can see a preview of an open document by hovering the cursor over the tab. In the same area, you can cascade, arrange, or list the open documents as shown in the following image.

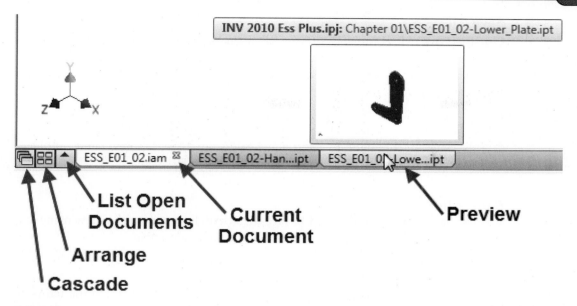

FIGURE 1.14

Save Options

There are three options on the Inventor Application button for saving your files: Save, Save Copy As, and Save All as shown in the following image.

FIGURE 1.15

Save

The Save command saves the current document with the same name and to the location where you created it. If this is the first time that a new file is saved, you are prompted for a file name and file location.

To run the Save command, click the Save icon on the Quick Access Toolbar, use the shortcut keys CTRL + S, or click Save on the File menu.	**TIP**

Save All
Use the Save All command to save all open documents and their dependents. The files are saved with the same name to the location where you created them. The first time that a new file is saved, you will be prompted for a file name and file location.

Save As
Use the Save As command to save the active document with a new name and location, if required. A new file is created and is made active.

Save Copy As
Use the Save Copy As command to save the active document with a new name and location, if required. A new file is created but is not made active.

Save Copy As Template
Use the Save Copy As Template command to save the current file to the template folder. New files can be based on the template file. Templates can be saved in the existing folders, or you can create a subdirectory in the *Autodesk\Inventor (version number)\templates* directory, and add a file to it. A new template tab, with the same name as the subdirectory, is created automatically when a file is added to the new folder.

Pack and Go
Use the Pack and Go command to copy all the files that are used to create the current file to a specified location.

Save Reminder
You can have Inventor remind you to save a file. Inventor will NOT automatically save the file. After a predetermined amount of time has expired without saving the file, a notification bubble appears in the upper right corner of the screen as shown in the following image on the left. The time can be adjusted via the Application Options > Save tab as shown in the following image on the right. The time can be adjusted from 1 minute to 9999 minutes, or uncheck this option to turn off the notification.

FIGURE 1.16

APPLICATION OPTIONS

Autodesk Inventor can be customized to your preferences. On the Inventor Application Menu, click Options, or from the Tools tab click Application Options, to open the Options dialog box as shown in the following image. You set options on each of the tabs to control specific actions in the Autodesk Inventor software. The application options affect all Inventor documents that are open or will be created. Each section is covered in more detail in the pertinent sections throughout this book. For more information about application options, see the Help system.

FIGURE 1.17

General
Set general options for how Autodesk Inventor operates.

Save
Set how files are saved.

File
Set where files are located.

Colors
Change the color scheme and color of the background on your screen. Determine if reflections and textures will be displayed.

Display
Adjust how parts look. Your video card and your requirements affect the appearance of parts on your screen. Experiment with different settings to achieve maximum video performance.

Hardware
Adjust the interaction between your video card and the Autodesk Inventor software. The software is dependent upon your video card. Take time to make sure that you are running a supported video card and the recommended video drivers. If you experience video-related issues, experiment with the options on the Hardware tab. For more information about video drivers, click Graphics Drivers on the Help menu.

Prompts
Modify the response given to messages that are displayed.

Drawing
Specify the way that drawings are created and displayed.

Notebook
Specify how the Engineer's Notebook is displayed.

Sketch
Modify how sketch data is created and displayed.

Part
Change how parts are created.

iFeature
Adjust where iFeatures data is stored.

Assembly
Specify how assemblies are controlled and behave.

Content Center
Specify the preferences for using the Content Center.

EXERCISE 1-1: USER INTERFACE

In this exercise, you change the user interface by moving the Ribbon, Quick Access Toolbar, and add and remove commands to the Quick Access Toolbar.

1. Click the New command, click the English tab, and then double-click Standard (in).ipt.

2. Move the Ribbon to different locations. Move the cursor anywhere over the Ribbon and right-click, and from the menu click Docking Positions and click the different options.
3. Undock the Ribbon, right-click on the Ribbon, and from the menu click Undock Ribbon. Move the Ribbon to different locations.
4. Move the Ribbon back to its original top position, right-click on the Ribbon, and from the menu click Docking Positions > Top as shown in the following image.

FIGURE 1.18

5. Change the appearance of the Ribbon, right-click on the Ribbon, and from the menu click Ribbon Appearance. Try the different options to change the Ribbon's appearance.
6. Reset the Ribbon back to its original state by clicking Reset Ribbon from the same menu.

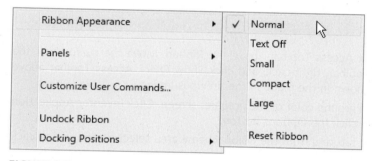

FIGURE 1.19

7. Add a command to the Quick Access Toolbar. Move the cursor over the Line command in the Draw panel, right-click, and click Add to Quick Access Toolbar.

FIGURE 1.20

8. Move the Quick Access Toolbar below the Ribbon. Move the cursor over the Quick Access Toolbar, right-click, and click Show Quick Access Toolbar below the Ribbon.

FIGURE 1.21

9. Remove the Line command from the Quick Access Toolbar. Move the cursor over the Line command in the Quick Access Toolbar, right-click, and click Remove from Quick Access Toolbar.

FIGURE 1.22

10. Move the Quick Access Toolbar above the Ribbon. Move the cursor over the Quick Access Toolbar, right-click, and click Show Quick Access Toolbar above the Ribbon as shown in the menu in the previous image.
11. Change the background color of the graphics screen. Click Inventor Application Menu, and then click the Options button.
12. Click the Colors tab, and from the Color scheme area select an option and click Apply to see the change.
13. Experiment with the background options.
14. Experiment changing the colors of the icons. In the Color Theme area, click the Amber option and click Apply. Notice the color of the icons change.
15. Change the icons color back to Cobalt as shown in the following image and click OK.

FIGURE 1.23

16. As you work with Inventor, adjust the user interface to meet your requirements.
17. Close the file. Do not save changes. End of exercise.

Command Entry

There are several methods to issue commands in Autodesk Inventor. In the following sections, you will learn how to start a command. There are no right or wrong methods for starting a command, and with experience, you will develop your own preference.

To stop a command, either press the ESC key, right-click, and click Done from the menu or select another icon.

Panels and Tooltips

In the last section, you learned how to control the appearance of the ribbon. The main function of the Ribbon is to hold the commands in a logical fashion, which is done by dividing the commands into panels. To start a command from a panel, move the cursor over the desired icon, and a command tip appears with the name of the command. You can control the tooltip from the Application Options under the General tab as shown in the following image on the left. The first level tooltip displays an abbreviated command description as shown in the following image in the middle. If the cursor hovers over the icon longer, a more detailed command description will appear as shown in the following image on the right.

FIGURE 1.24

Some of the icons in the Panel have a small down arrow in the right side. Select the arrow to see additional commands. To activate a command, move the cursor over a command icon and click. The command that is selected will appear first in the list replacing the previous command.

FIGURE 1.25

Marking Menus

Autodesk Inventor also uses marking menus. These are text menus that pop up when you press the right mouse button. The shortcut menus are context sensitive and contain options that are relevant to the current task. The following image on the left shows the marking menu that appears while in the Line command. The top section lists possible commands, while the bottom menu displays options and commands to use while in the command. As you gain experience with Inventor you can start a command from the top portion of the marking menu by right-clicking and moving the cursor (before the menu appears) in the direction of the command and release the mouse button as shown in the following image on the right. This technique is also referred to as gesture behavior.

FIGURE 1.26

Autodesk Inventor Shortcut Keys

Autodesk Inventor has keystrokes called shortcut keys that are preprogrammed. While in a command, the tooltip displays the shortcut key in parenthesis if a shortcut key is available. To start a command via a shortcut key, press the desired preprogrammed key(s). The keys can be reprogrammed by clicking the Tools tab and click Customize.

REPEAT LAST COMMAND

To restart a command without reselecting the command in a panel bar, either press ENTER or the spacebar, or right-click and click the top entry in the menu, Repeat "the last command." The following image shows the Line command being restarted.

FIGURE 1.27

Minimize Dialog Boxes

You can control whether an individual dialog is minimized. To minimize a dialog box, click the up arrow near the bottom of the dialog box. The dialog box then displays only the horizontal title bar as shown in the image on the right. To maximize the dialog box, click the down arrow in the minimized dialog box as shown in the following image on the right. This option is set for each dialog box.

FIGURE 1.28

Undo and Redo

You may want to undo an action that you just performed, or undo an undo. The Undo command backs up Autodesk Inventor one function at a time. If you undo too far, you can use the Redo command to move forward one step at a time. The Zoom, Orbit, and Pan commands do not affect the Undo and Redo commands. To start the commands, select the command from the Quick Access Toolbar as shown in the following image. The Undo command is to the left, and the Redo command is to the right. The shortcut keys are CTRL + Z for Undo and CTRL + Y for Redo.

NOTE To set the Undo file size allocation, click Tools tab > Application Options. On the General tab of the Options dialog box, change the Maximum size of Undo file (MB).

FIGURE 1.29

HELP SYSTEM

The Help system in Autodesk Inventor goes beyond basic command definition by offering assistance while you design. The commands in the Information Toolbar on the top-right corner of the screen will assist you while you design. To get help on a topic, enter a keyword in the area entitled "Type a keyword or phrase." To see the other help mechanisms that make up the Help System, click the drop arrow next to the question mark as shown in the following image.

FIGURE 1.30

Other options to access the Help system include the following methods:

- Press the F1 key, and the Help system assists you with the active operation.
- Click an option on the Help menu.
- Click a Help option on the Information Toolbar.
- In any dialog box, click the ? icon.
- Click an option on the Get Started tab > Learn about Inventor panel.
- Click on How To on a shortcut menu initiated within an active command.

PROJECTS IN AUTODESK INVENTOR

Almost every design that you create in Autodesk Inventor involves more than a single file. Each part, assembly, presentation, and drawing created is stored in a separate file. Each of these files has its own unique file extension. There are many times when a design will reference other files. An assembly file, for example, will reference a number of individual part files and/or additional subassemblies. When you open the parent or top-level assembly, it must contain information that allows Autodesk Inventor to locate each of the referenced files. Autodesk Inventor uses a project file to organize and manage these file-location relationships. There is no limit to the number of projects you can create, but only one project can be active at any given time.

You can structure the file locations for a design project in many ways. A single-person design shop has different needs from a large manufacturing company or a design team with multiple designers working on the same project. In addition to project files, Autodesk Inventor includes a program called Autodesk Vault that controls basic check-out and check-in file-reservation mechanisms; these control file access for multiuser design teams. Autodesk Inventor always has a project named Default. Specifically, if all the files defining a design are located in a single folder, or in a folder tree where each referenced part is located with its parent or in a subfolder underneath the parent, the Default project may be all that is required.

NOTE: It is recommended that files in different folders never have the same name to avoid the possibility of Autodesk Inventor resolving a reference to a file of the same name but in a different folder.

Project Setup

To reduce the possibility of file resolution problems later in the design process, always plan your project folder structure before you start a design. A typical project might consist of parts and assemblies unique to the project; standard components that are unique to your company; and off-the-shelf components such as fasteners, fittings, or electrical components.

Project File Search Options

Before you create a project, you need to understand how Autodesk Inventor stores cross-file reference information and how it resolves that information to find the referenced file. Autodesk Inventor stores the file name, a subfolder path (if present) to the file, and a library name (optionally) as the three fundamental pieces of information about the referenced file.

When you use the Default project file, the subfolder path is located relative to the folder containing the referencing file. It may be empty or may go deeper in the subfolder hierarchy, but it can never be located at a level above the parent folder.

When you create a project file, you do not need to add subfolder as search paths. The subfolder(s) path is automatically searched and does not need to be added to the project file.

Creating Projects

To create a new project or edit an existing project, use the Autodesk Inventor Project File Editor. The Project File Editor displays a list of shortcuts to previously active projects. A project file has an *.ipj* file extension and typically is stored in the home folder for the design-specific documents, while a shortcut to the project file is stored in the Projects Folder. The Projects Folder is specified on the Files tab of the Options dialog box as shown in the following image. All projects with a shortcut in the Projects Folder are listed in the top pane of the Project File Editor.

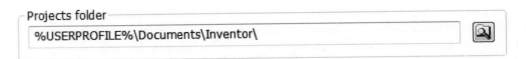

FIGURE 1.31

You create or edit a project file by clicking the Project button in the New or Open dialog box or by clicking Get Started tab > Launch panel > Projects as shown in the following image on the left, or from the Inventor Application Menu click Manage and click Projects as shown in the following image on the right.

FIGURE 1.32

The Projects dialog box will appear as shown in the following image. The Projects dialog box is divided into two panes. The top pane lists shortcuts to the project files that have been active previously. Double-click on a project's name to make it the active project. All Inventor files must be closed before making a project current. Only one project file can be active in Autodesk Inventor at a time. The bottom portion reflects information about the project selected in the top pane. If a project file already exists, click on the Browse button on the bottom of the dialog box, then navigate to, and select the project file. The bottom pane of the dialog box lists information about the highlighted project.

When defining a path to a folder on a network, it is better to define a Universal Naming Convention (UNC) path starting with the server name (\\Server\...) and not to use shared (mapped) network drives.	**NOTE**

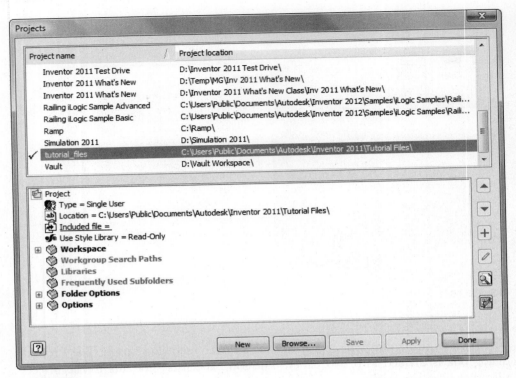

FIGURE 1.33

To create a new project, follow these steps:

1. In the Projects dialog box, click the New button at the bottom to initiate the Inventor project wizard.
2. In the Inventor project wizard, follow the prompts to the following questions.

What Type of Project Are You Creating?
If Autodesk Vault is installed, you will be prompted to create a New Vault project or a New Single-User project. If Autodesk Vault is not installed, only a New Single-User project type will appear in the list.

New Vault Project
This project type is used with Autodesk Vault and is not available until you install Autodesk Vault.

New Single-User Project
This is the default project type, which is used when only one user will reference Autodesk Inventor files. It creates one workspace where Autodesk Inventor files are stored and any needed library location(s), and it sets the Project Type to Single User. No workgroup is defined but can be defined later.

The next section covers the steps for creating a new single-user project. For more information on projects, consult the online Help system.

Creating a New Single-User Project

Click on New Single User Project as shown in the following image. If Autodesk Vault is not installed, New Single User Project will be selected for you.

FIGURE 1.34

Click the Next button, and specify the project file name and location on the second page of the Inventor project wizard as shown in the following image.

Name

Enter a descriptive project file name in the Name field. The project file will use this name with an *.ipj* file extension.

FIGURE 1.35

Project (Workspace) Folder

This specifies the path to the home or top-level folder for the project. You can accept the suggested path, enter a path, or click the Browse button (…) to manually locate the path. The default home folder is a subfolder under My Documents, or wherever you last browsed, that is named to match the project file name.

Project File to Be Created

The full path name of the project file is displayed below the Location field.

> **NOTE**
> You can specify the home or top-level folder as the location of your workspace, but it is preferable for the project file (.*ipj*) to be the only Autodesk Inventor file stored in the home folder. This makes it easier to create other project locations, such as library and workspace folders, as subfolders without creating nesting situations that can lead to confusion.

Click the Next button at the bottom of the Inventor project wizard, and specify the project library search paths.

You can add library search paths from existing project files to this new project file. The library search paths from every project with a shortcut in your Projects Folder are listed on the left in the Inventor project wizard dialog box as shown in the following image. You can add and remove libraries from the New Project area by clicking on their names in either the All Projects area or the New Project area and then clicking the arrows in the middle of the dialog box. The libraries listed by default in the All Projects list will match those in the project file that you selected prior to starting the New Project process.

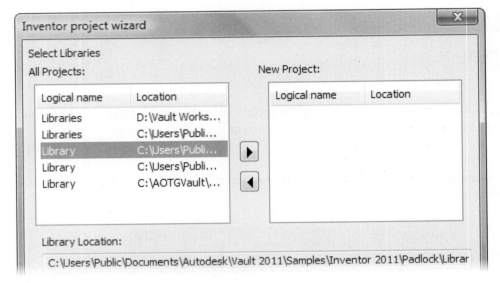

FIGURE 1.36

Click the Finish button to create the project. If a new directory will be created, click OK in the Inventor Project Editor dialog box. The new project will appear in the Open dialog box. Double-click on a project's name in the Projects dialog box to make it the active project. A checkmark will appear to the left of the active project as shown in the following image.

FIGURE 1.37

EXERCISE 1-2: PROJECTS

In this exercise, you create a project file for a single-user project, open existing files, delete the project file, and make an existing project file current.

1. Prior to creating a new project file, close all Inventor files and ensure that the exercise files have been installed.
2. From the Get Started tab > Launch panel click Projects.
3. You now create a new project file. Click the New button at the bottom of the Projects dialog box.
4. Click New Single User Project, and click Next.
5. In the next dialog box of the Inventor project wizard, enter **Inventor Exercises** in the Name field.
6. Click the Browse button (...) to the right of the folder location and name of the Project (Workspace) Folder field.
7. Browse to and select the folder *C:\Inv 2012 School* folder as shown in the following image. The project file (*.ipj*) will be placed in this folder. The selected folder is the top-level folder for the project. It is a good practice to place project component files such as parts, assemblies, and drawings in folders below the top-level folder, but not in the top-level folder.

FIGURE 1.38

8. Click Next to add a library search path. The New Project list on the right side should be blank.
9. Click Finish. The new project file is highlighted in the upper pane of the Project File Editor. The search paths for the new project are listed in the lower pane.
10. Activate the new project by double-clicking on the project named **Inventor Exercises** in the upper pane of the Project File Editor.
11. Expand Workspace and notice that the Workspace search path is listed as a period "." as shown in the following image. The "." denotes that the workspace location is relative to the location where the project file is saved.

FIGURE 1.39

12. To make it easier to use subfolders below the workspace, you can add them to the Frequently Used Subfolders. Right-click the Frequently Used Subfolders option, and click Add Paths from Directory as shown in the following image. Then navigate to and select the *C:\Inv 2012 School* folder. Click OK to add the subfolders.

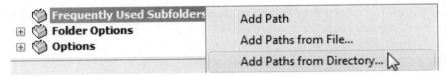

FIGURE 1.40

13. Click Save, and then click Done.
14. Click the Open icon on the Quick Access Toolbar. Notice that the 7 chapters appear in the Frequently Used Subfolders area in the upper-left corner of the dialog box as shown in the following image.

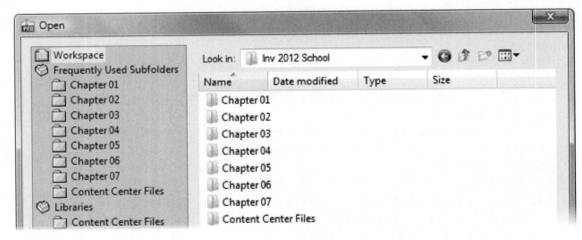

FIGURE 1.41

15. Click any of the chapter folders under the Frequently Used Subfolders, and then open any of the Inventor files that exist in the folder. Notice that the Look in location changes to match the subfolder name.
16. **You must change the active project file to successfully complete the remaining exercises. Close all files currently open in Autodesk Inventor.**

17. Click Inventor Application Menu then click Manage > Projects.
18. To add an existing project file to the current list, click the Browse ... button at the bottom of the dialog box. Navigate to and select *C:\Inv 2012 School\Inv 2012 School.ipj*. This project should now be the current project. This project will be used for the remaining exercises. Expand the Frequently Used Subfolders section to verify that each chapter has its own subfolder.
19. Right-click on the project **Inventor Exercises** in the upper pane and select delete as shown in the following image.

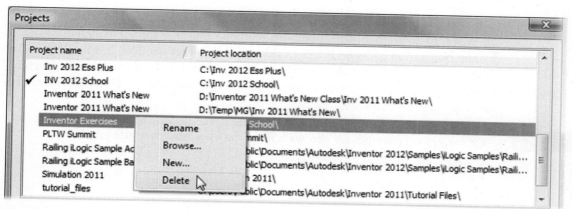

FIGURE 1.42

20. End of exercise.

VIEWPOINT OPTIONS

When you work on a 2D sketch, the default view is looking straight down at the XY plane, which is often referred to as a plan view. When you work in 3D, it is helpful to view objects from a different viewpoint and to zoom in and out or pan the objects on the screen. The next section guides you through the most common methods for viewing objects from different perspectives and viewpoints. As you use these commands, the physical objects remain unmoved. Your perspective or viewpoint of the objects is what creates the perceived movement of the part. If you are performing an operation while a viewing command is issued, the operation resumes after the transition to the new view is completed.

Home (Isometric) View

Change to an isometric viewpoint by pressing the F6 key, click the home icon above the ViewCube as shown in the following image on the left, or by right-clicking in the graphics window and then selecting Home View from the menu as shown in the following image on the right. The view on the screen transitions to a predetermined home view. You can redefine the home view with a ViewCube option. ViewCube is explained later in this section.

FIGURE 1.43

Navigation Bar

The Navigation bar, as shown in the following image, contains commands that will allow you to zoom, pan, and rotate the geometry on the screen. The default location for the Navigation bar is on the right side of the graphics window, but it can be repositioned. Commands are also available below the displayed commands. When commands are selected, they will become the top level command. Descriptions of the viewing commands follow.

NOTE The navigation commands are also available on the View tab > **Navigate** panel.

FIGURE 1.44

Navigation Wheel

Click the Navigation Wheel icon to turn on the steering wheel. Click one of the eight options and that option will be active.

    Orbit – Rotates the viewpoint.

    Zoom – Zooms in and zooms out.

    Rewind – Changes to a previous view. A series of images appears representing the previous views. Move the cursor over the image that represents the view to restore.

    Pan – Moves the view to a new location.

    Center – Centers the view based on the position of the cursor over the wheel.

    Walk – Swivels the viewpoint.

    Up/Down – Changes the viewpoint vertically.

    Look – Moves the view without rotating the viewpoint.

    Click the down arrow below the Navigation Wheel icon to change the look of the Navigation wheel.

Pan

Moves the view to a new location. Issue the Pan View command, or select the F2 key. Press and hold the left mouse button, and the screen moves in the same direction that the cursor moves. If you have a mouse with a wheel, hold down the wheel, and the screen moves in the same direction that the cursor moves.

Zoom All

Maximizes the screen with all parts that are in the current file. The screen transitions to the new view. Other commands under Zoom All command are Zoom, Zoom Window, and Zoom Selected.

Zoom. Zooms in or out from the parts. Issue the Zoom In-Out command, or press the F3 key. Then, in the graphics window, press and hold the left mouse key. Move the mouse toward you to make the parts appear larger, and away from you to make the parts appear smaller. If you have a mouse with a wheel, roll the wheel toward you, and the parts appear larger; roll the wheel away from you, and the parts appear smaller.

Zoom Window. Zooms in on an area that is designated by two points. Issue the Zoom Window command and select the first point. With the mouse button depressed, move the cursor to the second point. A rectangle representing the window appears. When the correct window is displayed on the screen, release the mouse button, and the view transitions to it.

Zoom Selected. Fills the screen with the maximum size of a selected face, faces, or a part. Either select the face or faces and then issue the Zoom Selected command, or press the END key, or launch the Zoom Selected command, or press the END key. Then select the face or faces to which you wish to zoom. Use the Select Other command to select a part to zoom to.

Orbit

Rotates your viewpoint dynamically. Issue the Orbit command; a circular image with lines at the quadrants and center appears. To rotate your viewpoint, click a point inside the circle, and keep the mouse button pressed as you move the cursor. Your viewpoint rotates in the direction of the cursor movement. When you release the mouse button, the viewpoint stops rotating. To accept the view orientation, either press the ESC key or right-click and select Done from the menu. Click the outside of the circle to rotate the viewpoint about the center of the circle. To rotate the viewpoint about the vertical axis, click one of the horizontal lines on the circle and, with the mouse button pressed, move the cursor sideways. To rotate the viewpoint about the horizontal axis, click one of the vertical lines on the circle and, with the mouse button pressed, move the cursor upward or downward. The Constrained Orbit command is available below the Orbit command.

Constrained Orbit. Use the Constrained Orbit command to rotate the model about the horizontal screen axes like a turntable. Click one of the horizontal lines on the circle and, with the mouse button pressed, move the cursor sideways. The model is rotated about the model space center point set in the Navigation Wheel CENTER command.

View Face

Changes your viewpoint so that you are looking perpendicular to a plane or rotates the screen viewpoint to be horizontal to an edge. Issue the View Face command or press the PAGE UP key; then select a plane or edge. The View Face command can also be issued by selecting a plane or edge, right-clicking while the cursor is in the graphics window, and selecting Look At from the menu.

Customize the Navigation Bar

To modify the Navigation bar, click the down arrow on the bottom of the Navigation Toolbar as shown in the following image. Click a command to add or remove it from the Navigation bar. You can also reposition the toolbar by clicking the options under Docking position.

FIGURE 1.45

ViewCube

The ViewCube allows you to quickly change the viewpoint of the screen. With the ViewCube turned on, move the cursor over the ViewCube and the opacity becomes 100%. Move the cursor over the home in the ViewCube to return to the default home (isometric) view as shown in the following image on the left.

FIGURE 1.46

Change the viewpoint by using one of the following techniques.

Isometric
Change to a different isometric view by clicking a corner on the ViewCube as shown in the following image, second from the left.

Face
Change the viewpoint so it looks directly at a plane by clicking a plane on the ViewCube as shown in the following middle image.

Rotate in 90 Degree Increments
When looking at the plane of a ViewCube, you can rotate the view 90 degrees by clicking one of the arcs with arrows or one of the four inside facing triangles as shown in the following image second from the right.

Edge
You can also orientate the viewpoint to an edge. Click an edge on the ViewCube as shown in the following image on the right.

FIGURE 1.47

Dynamic Rotate
To dynamically rotate your viewpoint, click on the ViewCube and keep the mouse button pressed as you move the cursor. Your viewpoint rotates in the direction of the cursor movement. When you release the mouse button, the viewpoint stops rotating. When the viewpoint does not match a defined viewpoint, the ViewCube edges appear in dashed lines from one of the corners as shown in the following image. To smooth the rotation, right-click on the ViewCube and click Options > and uncheck Snap to closest view.

FIGURE 1.48

ViewCube Options

To change the ViewCube's options, move the cursor over the ViewCube and click the down arrow as shown in the following image. From the menu you can change the following:

FIGURE 1.49

Go Home
Change the viewpoint to the default Home View.

Orthographic
Set the display viewpoint to be parallel; the lines are projected perpendicular to the plane of projection.

Perspective
Set the display viewpoint to be perspective; the geometry on the screen converges to a vanishing point similar to the way the human eye sees.

Perspective with Ortho Faces
Set the display to perspective except when looking directly at a plane of the ViewCube.

Lock to Current Selection
The center and distance does not change regardless of whether objects are selected or deselected.

Set Current View as Home – Fixed Distance
Defines the current view as the default home view. The view will return to the same size and orientation as when it was set.

Set Current View as Home – Fit to View
Defines the current view as the default home view. The view size will always display all of the parts, but have the same orientation as when it was set.

Set Current View as – Front or Top View
Defines the current view as the front or top view. The view does not need to be a plane on the ViewCube.

Reset Front
Resets the front view to the default setting.

Options
Opens the ViewCube Options dialog box.

Help Topics
Launches the Help system and displays the topic on the ViewCube.

Dynamic Rotation
Rotates viewpoint dynamically. While working, press and hold down the F4 key. The circular image appears with lines at the quadrants and center. With the F4 key still depressed, rotate the viewpoint. When you finish rotating the viewpoint, release the F4 key. If you are performing an operation while the F4 key is pressed, that operation will resume after you release the F4 key.

Another option to quickly rotate your viewpoint is to hold down the SHIFT key + Wheel (middle mouse button). Once the rotate glyph appears on the screen, you can release the SHIFT key and, while holding down the wheel, move the mouse, and your viewpoint will rotate. With this option, no other options are available. The model(s) are rotated about the center of the model(s). Release the wheel to complete the operation.

Appearance Options
While working with a 3D model, you can adjust the perspective and shading option and control the shadows. To change the appearance, click the View tab > Appearance panel as shown in the following image, or you can add Visual Styles to the Navigation bar.

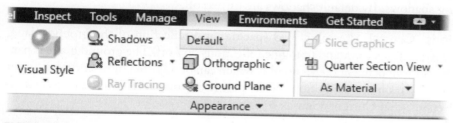

FIGURE 1.50

Visual Style
Set the appearance of model faces and edges in the graphics window. There are ten different styles to choose from as shown in the following image. The styles do not change the model properties; they just change how the model is displayed. Combine visual styles with the other appearance options to display your model(s) differently.

FIGURE 1.51

Shadows. To give your model a realistic look, you can choose different options for displaying shadows. By default shadows are turned off.

Reflections. Controls the color, grid spacing, and reflection of the ground plane. Click the Reflection command to toggle reflections on / off. The ground plane is parallel to the Bottom plane of the view cube, which can be adjusted by orienting the view so the view is parallel to the bottom of the part (that you want). Then right-click on the view cube > Set Current View as > Top.

Ray Tracing. When the Realistic visual style option is selected, the Ray Tracing option will be turned on. Ray tracing gives the images a more realistic effect by simulating how the light reflects off the part(s). Ray tracing will add to the time it takes to render the image and should only be done when a very realistic image is required.

Lights. Select or set different lighting options. From under the Lights command click the Settings option. From within the dialog box you can control the lights direction, color, add lights and control brightness and ambience. Also from this dialog box you can turn on Image Lighting, which allows you to set a background image and set the scale of the image. With a scene image set you can rotate your viewpoint 360 degrees. To move the image, select the Ground Plane Settings option, change the Position & Size option to Manual adjustment and a triad will appear. Select the arrows on the triad to move the image.

Orthographic. Sets the viewpoint to be orthographic or perspective. When set to orthographic the viewpoint is parallel, meaning that lines are projected perpendicular to the plane of projection. For clarity, this book shows all the images in the orthographic viewpoint. When set to perspective the geometry on the screen converges to a vanishing point similar to the way the human eye sees.

Color. Sets the color of the current part. The color can also be set from the Quick Access toolbar.

EXERCISE 1-3: VIEWING A MODEL

In this exercise, you use View Manipulation commands that make it easier to work on your designs.

1. Open *C:\Inv 2012 School\Chapter 01\ESS_E01_03.iam*; the file contains a model of a clamp.

> **TIP** Click the Chapter 01 subfolder from the Frequently Used Subfolders area, and then click the file in the file area as shown in the following image.

FIGURE 1.52

2. Move the cursor over the bottom clamp grips and spin the wheel toward you. Inventor will zoom in/out to the location of the cursor. This will fill the screen as shown in the following image.

FIGURE 1.53

3. Press F5 to return to the previous view.
4. Use the Orbit command to rotate the viewpoint. From the Navigation bar, click the Orbit command. The 3D Rotate symbol appears in the window as shown in the following image.

FIGURE 1.54

5. Move the cursor inside the circle of the 3D Rotate symbol, noting the cursor display.
6. Click and drag the cursor to rotate the model.
7. To return to the Home View, press F6.
8. Place the cursor on one of the horizontal handles of the 3D Rotate symbol, noting the cursor display.
9. Drag the cursor right or left to rotate the viewpoint about the Y axis. As you rotate the viewpoint, you will see the bottom of the assembly.
10. To again rotate the viewpoint about the vertical axis like a turntable, click the Constrained Orbit command below the Orbit command on the Navigation bar. Click one of the horizontal lines on the circle. With the mouse button pressed, move the cursor sideways. The view will rotate like a turntable; notice that you do not see the bottom of the assembly.
11. Return to the home view and click the Home symbol above the ViewCube.

TIP — While you are performing another process, you can click on the viewing commands, spin the wheel to zoom, press and hold the wheel to pan, or press and hold F4 to rotate the view.

12. To change the viewpoint to predetermined isometric views, or directly at a plane, use the ViewCube. Click the Top-Front left corner of the ViewCube as shown in the following image on the left. Continue clicking the other corners of the ViewCube as shown in the following middle two images.
13. Change to the front view by clicking the Front plane on the ViewCube as shown in the following image on the right.

FIGURE 1.55

14. Rotate the view 90 degrees by clicking the arc with arrow in the ViewCube as shown in the following image on the left.
15. To look at the Top view that is also rotated by 90 degrees, click the left arrow as shown in the following image on the right.

FIGURE 1.56

16. Practice changing views by clicking the corners, faces, and edges of the ViewCube.
17. Click the Navigation Wheel icon on the Navigation bar.
18. Click the different options shown in the following image on the Navigation Wheel and experiment with them.

FIGURE 1.57

19. Move the cursor into a blank area in the graphics window, right-click, and click Home View from the menu.
20. Next you will apply different Visual Styles. Click the View tab > Appearance panel > under Visual Style; experiment by selecting different styles.
21. Change the visual style to Shaded.
22. Next you will apply different Shadows. Click the View tab > Appearance panel > under Shadows and experiment by turning the different shadows on and off.
23. Change the Shadow option to All Shadows.
24. Turn on reflections. From the View tab > Appearance panel click Reflections.

25. Next change to a perspective view. From the View tab > Appearance panel click the arrow beside Orthographic and click Perspective.
26. To change the perspective, press CTRL + Shift, and spin the wheel on the mouse.
27. Change back to an orthographic view. From the View tab > Appearance panel, click the arrow beside Perspective and click Orthographic, and your screen should resemble the following image.

FIGURE 1.58

28. Apply different lights. From the View tab > Appearance panel, click the arrow beside Default and select different lighting configurations.
29. In the Style and Standard Editor dialog box from the Lighting option make the following changes as shown in the following image.
    - Click the Use Image Lighting option to turn it on.
    - Click the Display Scene Image option to turn it on.
    - From the Source drop list select Empty Lab.
    - Adjust the image Scale to 25% and move the slider under the Scale option.
    - To complete the operation click Done and then click Yes to save the changes. Note that you could have created a new Lighting Style.

FIGURE 1.59

30. Adjust the location of the image by clicking the down arrow to the right of the Ground Plane command and click Settings. Change the Position & Size option to Manual Adjust, in the graphics window click and drag the arrows that appear. When done click OK.

FIGURE 1.60

31. Change the Visual Style to Realistic and experiment with the Ray Tracing options.
32. Continue to practice applying different appearance options.
33. Close the file. Do not save changes model. End of exercise.

CHECKING YOUR SKILLS

Use these questions to test your knowledge of the material covered in this chapter.

1. Explain the reasons why a project file is used.
2. True ___ False ___ Only one project can be active at any time.
3. True ___ False ___ Subfolders need to be added to the project file in order for the files to be found.
4. True ___ False ___ Autodesk Inventor stores the part, assembly information, and related drawing views in the same file.
5. True ___ False ___ Press and hold down the F4 key to dynamically rotate the viewpoint.
6. True ___ False ___ The Save Copy As command saves the active document with a new name, and then makes it current.
7. List four ways to access the Help system.
8. Explain how to change the location of the Ribbon.
9. True ___ False ___ Commands can be added to the Quick Access toolbar.
10. True ___ False ___ Visual styles change the physical properties of the model.
11. True ___ False ___ Ray Tracing is used to get an accurate mass property.
12. True ___ False ___ The settings in the Application Options are global and affect all open and new Inventor files.
13. Explain how to move the Quick Access toolbar below the ribbon.
14. Explain how to start a command from the marking menu without clicking a specific command.
15. True ___ False ___ By default Inventor will automatically save the current file every 10 minutes.

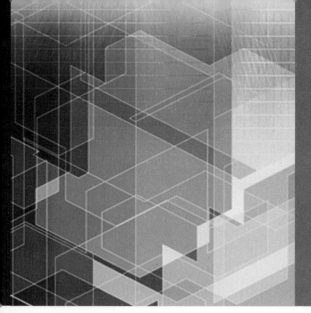

CHAPTER 2

Sketching, Constraining, and Dimensioning

INTRODUCTION

Most 3D parts in Autodesk Inventor start from a 2D sketch. This chapter first provides a look at the application options for sketching and part creation. It then covers the three steps in creating a 2D parametric sketch: sketching a rough 2D outline of a part, applying geometric constraints, and then adding parametric dimensions.

OBJECTIVES

After completing this chapter, you will be able to do the following:

- Change the part and sketch options as needed
- Sketch an outline of a part
- Create geometric constraints
- Use construction geometry to help constrain sketches
- Dimension a sketch
- Create dimensions using the automatic dimensioning command
- Change a dimension's value in a sketch

PART AND SKETCH APPLICATION OPTIONS

Before you start a new part, examine the part and sketch options in Autodesk Inventor that will affect how the part file will be created and how the sketching environment will look and act. While learning Autodesk Inventor, refer back to these option settings to determine which ones work best for you—there are no right or wrong settings.

Part Options

You can customize Autodesk Inventor Part options to your preferences. Click Inventor Application Menu > Options button, and click on the Part tab, as shown in the following image. These settings are global—they will affect all active and new Autodesk Inventor documents.

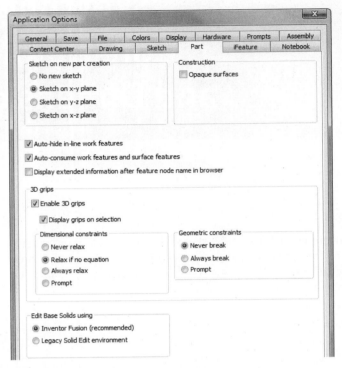

FIGURE 2.1

Sketch Options

Autodesk Inventor sketching options can be customized to your preferences. Click Inventor Application Menu > Options, and then click on the Sketch tab, as displayed in the following image. A description of a few of the options follow. For more information about the Part Application Options, consult the help system. These settings are global, and all of them affect currently active Autodesk documents and Autodesk Inventor documents you open in the future.

FIGURE 2.2

Edit Dimension When Created
When checked, edits the values of a dimension in the Edit Dimension dialog box immediately after you position the dimension. It is recommended to turn this setting on.

Autoproject Edges During Curve Creation
When checked, and while sketching, place the cursor over an object and it will be projected onto the current sketch. You can also toggle Autoproject on and off while sketching by right-clicking and selecting Autoproject from the menu.

Autoproject Edges for Sketch Creation and Edit
When checked, automatically projects all of the edges that define that plane onto the sketch plane as reference geometry when you create a new sketch.

Look at Sketch Plane on Sketch Creation
When checked, automatically changes the view orientation to look directly at the new or active sketch.

Autoproject Part Origin on Sketch Create
When checked, the part's origin point will automatically be projected when a new sketch is created. It is recommended to keep this setting on.

UNITS

Autodesk Inventor uses a default unit of measurement for every part and assembly file. The default unit is set from the template file from which you created the part or assembly file. When specifying numbers in dialog boxes with no unit, the default unit will be used. You can change the default unit in the active part or assembly document by clicking Tools tab > Document Settings button and selecting the Units tab as shown in the following image. The unit system values change for all of the existing values in that file.

NOTE In a *drawing* file, the appearance of dimensions is controlled by dimension styles. Drawing settings will be covered in Chapter 5.

FIGURE 2.3

You can override the default unit for any value by entering the desired unit. If you were working in a metric file whose unit is set to mm, for example, and you placed a horizontal dimension whose default value was 50.8 mm, you could enter 2 in. Dimensions appear on the screen in the default units.

For the previous example, 50.8 mm would appear on the screen. When you edit a dimension, the overridden unit appears in the Edit Dimension dialog box as shown in the following image.

FIGURE 2.4

TEMPLATES

Each new file is created from a template. You can modify existing templates or add your own templates. As you work, make note of the changes that you make to each file. You then create a new template file or modify an existing file that contains all of the changes and save that file to your template directory, which by default in Windows XP is located at *C:\Program Files\Autodesk\Inventor 2012\templates* directory and in Windows Vista and Windows 7 is *C:\Users\Public\Public Documents\ Autodesk\ Inventor 2012\Templates* directory. You can also create a new subdirectory under the templates folder, and place any Autodesk Inventor file in this new directory. After adding an Inventor file the new tab will appear, and it will be available as a template.

You can use one of two methods to share template files among many users. You can modify the location of templates by clicking the Inventor Application Menu > Options button, clicking on the File tab, and modifying the Templates location as shown in the following image. The Templates location will need to be modified for each user who needs access to these common templates.

Default templates
%PUBLICDOCUMENTS%\Autodesk\Inventor 2012\Templates\

FIGURE 2.5

You can also set the Templates location in each project file. This method is useful for companies that need different templates files for each project. While editing a project file, change the Templates location in the Folder Options area. The following image shows the default location in Windows Vista. The Template location in the project file takes precedence over the Templates option in the Application Options, File tab.

Folder Options
Design Data (Styles, etc.) = [Default]
Templates = [Default]
Content Center C:\Users\Public\Documents\Autodesk\Inventor 2012\Templates\

FIGURE 2.6

> Template files have file extensions that are identical to other files of the same type, but they are located in the template directory. Template files should not be used as production files.

NOTE

CREATING A PART

The first step in creating a part is to start or create a new part file in an assembly. The first four chapters in this book deal with creating parts in a new part file, and Chapter 6 covers creating and documenting assemblies. You can use the following methods to create a new part file:

- Click New on the Inventor Application menu, and then click the *Standard.ipt* icon on the Default tab, as shown in the following image on the left. You can also click *Standard (unit).ipt* on one of the other tabs.
- Click the down arrow of the New icon, and select Part from the left side of the Quick Access toolbar as shown in the following image on the right. This creates a new part file based on the units that were selected when Autodesk Inventor was installed or on the *Standard.ipt* that exists in the Templates folder.
- Click the New icon in the Quick Launch area of the Open dialog box, and then click the *Standard.ipt* icon on the Default tab, as shown in the following image on the left, or click *Standard (unit).ipt* on one of the other tabs.

- Use the shortcut key, CTRL-N, and then click the *Standard.ipt* icon on the Default tab, as shown in the following image. You can also click *Standard (unit).ipt* on one of the other tabs.

After starting a new part file using one of the previous methods, Autodesk Inventor's screen will change to reflect the part environment.

> **NOTE** The units for the files located in the Default tab are based upon the units you selected when you installed Autodesk Inventor.

FIGURE 2.7

Sketches and Default Planes

Before you start sketching, you must have an active sketch on which to draw. A sketch is a plane on which 2D objects are sketched. You can use any planar part face or work plane to create a sketch. A sketch is automatically set by default when you create a new part file. You can change the default plane on which you will create the sketch by selecting the Inventor Application Menu > Options and clicking on the Part tab. Choose the sketch plane to which new parts should default.

Each time you create a new Autodesk Inventor, there are three planes (XY, YZ, and XZ), three axes (X, Y, and Z), and the center (origin) point at the intersection of the three planes. You can use these default planes to create an active sketch. By default, visibility is turned off to the planes, axes, and center point. To see the planes, axes, or center point, expand the Origin entry in the browser by clicking on the + to the left side of the text. You can then move the cursor over the names, and they will appear in the graphics area. The following image illustrates the default planes, axes, and center point in the graphics area shown in an isometric view with their visibility on. The browser shows the Origin menu expanded. To leave the visibility of the planes or axes on, right-click in the browser while the cursor is over the name, and select Visibility from the menu.

FIGURE 2.8

Origin 3D Indicator

When working in 3D, it is common to get your orientation turned around. By default in the lower left corner of the graphics screen, there is an XYZ axis indicator that shows the default (world) coordinate system as shown in the following image on the left. The direction of these planes and axes cannot be changed. The arrows are color coded:

Red arrow = X axis

Green arrow = Y axis

Blue arrow = Z axis

In the Application Options dialog box under the Display tab, you can turn the axis indicator and the axis labels on and off as shown in the following image on the right.

Origin 3D indicator
☑ Show Origin 3D indicator
☑ Show Origin XYZ axis labels

FIGURE 2.9

Autoproject Center (Origin) Point

To access another option that will automatically project the origin point for each new sketch of a part, click Tools tab > Application Options, click on the Sketch tab, and then check Autoproject part origin on sketch create as displayed in the following image. The origin point can be used to constrain a sketch to the 0, 0 point of the sketch. This option is on by default.

☑ Autoproject part origin on sketch create

FIGURE 2.10

New Sketch

Issue the Sketch command to create a new sketch on a planar part face or a work plane or to activate a nonactive sketch in the active part. To create a new sketch or make an existing sketch active, use one of these methods:

- Click the Model tab > Sketch panel > Create 2D Sketch as shown in the following image on the left. Then click a face, a work plane, or an existing sketch in the browser.
- Click a face, a work plane, or an existing sketch in the browser. Then click the Create 2D Sketch command on the Model tab.
- Press the S key (a keyboard shortcut) and click a face of a part, a work plane, or an existing sketch in the browser.
- Click a face of a part, a work plane, or an existing sketch in the browser, and then press the S key.
- While not in the middle of an operation, right-click in the graphics area, and select New Sketch from the marking menu. Then click a face, a work plane, or an existing sketch in the browser.
- While not in the middle of an operation, click a face of a part, a work plane, or an existing sketch in the browser. Then right-click in the graphics area, and select Create Sketch from the mini-toolbar as shown in the following image on the right.

FIGURE 2.11

By default after you have activated the sketch, the X- and Y-axes will align automatically to this plane, and you can begin to sketch.

> **NOTE** This book assumes that when you installed Autodesk Inventor, you selected mm as the default unit. If you selected inch as the default unit, then select the *Standard (mm).ipt* template file from the Metric tab. This book will use the XY plane as the default sketch plane.

STEP 1—SKETCH THE OUTLINE OF THE PART

As stated at the beginning of this chapter, 3D parts usually start with a 2D sketch of the outline shape of the part. You can create a sketch with lines, arcs, circles, splines, or any combination of these elements. The next section will cover sketching strategies, commands, and techniques.

Sketching Overview

When deciding what outline to start with, analyze how the finished shape will look. Look for the 2-dimensional shape that best describes the part. When looking for this outline, try to look for a flat 2-dimensional shape that can be extruded or revolved to create a shape that other features can be added to, to create the finished part. It is usually easier to sketch a 2-dimensional geometry than 3-dimensional geometry. As you gain modeling experience, you can reflect on how you created the model and think about other ways that you could have built it. There is usually more than one way to generate a given part.

When sketching, draw the geometry so that it is close to the desired shape and size—you do not need to be concerned about exact dimensional values. Even though Autodesk Inventor allows islands in the sketch (closed objects that lie within another closed object) it is NOT recommended to sketch islands (when you extrude a sketch, island(s) may become voids in the solid). A better method is to place features, which make editing a part easier. For example, instead of sketching a circle inside a rectangle, place a hole feature. A sketch can consist of multiple closed objects that are coincident.

The following guidelines will help you to successfully generate sketches:

- Select a 2-dimensional outline that best represents the part. The 2D outline will be used to create the base feature.
- Draw the geometry close to the finished size. If you want a 20 inch square, for example, do not draw a 200 inch square. Use dynamic input to define the size of the geometry. Dynamic input is covered in a later section in this chapter.
- Create the sketch proportional in size to the finished shape. When drawing the first object verify its size in the lower-right corner of status bar. Use this information as a guide.
- Draw the sketch so that it does not overlap. The geometry should start and end at a single point, just as the start and end points of a rectangle share the same point.
- Do not allow the sketch to have a gap; all of the connecting endpoints should be coincident.
- Keep the sketches simple. Leave out fillets and chamfers when possible. You can easily place them as features after the sketch turns into a solid. The simpler the sketch, the fewer the number of constraints and dimensions that will be required to constrain the model.

If you want to create a solid, the sketch must form a closed shape. If it is open, it can only be turned into a surface.

Sketching Tools

Before you start sketching the outline of the part, examine the 2D sketching commands that are available. By default, the 2D sketch commands appear on the Panel Bar with Display Text with Icons turned on (text descriptions shown). As you become more proficient with Autodesk Inventor, you can turn off the text next to some of the commands, as shown in the following image. Do this by right-clicking on the ribbon, and from the Appearance menu click Text Off. The most frequently used commands will be explained throughout this chapter. Consult the help system for information about the remaining commands.

FIGURE 2.12

Using the Sketch Tools

After starting a new part, a sketch will automatically be active so that you can now use the sketch commands to draw the shape of the part. To start sketching, issue the sketch command that you need, click a point in the graphics area, and follow the prompt on the status bar. The sections that follow will introduce techniques that you can use to create a sketch.

Dynamic Input

Dynamic input in the sketch environment makes a Heads-Up Display (HUD), which displays information near the cursor for many sketching commands that helps you keep your eyes on the screen. While using the Line, Circle, Arc, Rectangle, or Point commands, you can enter values in the input fields. You can toggle between the value input fields by pressing the TAB key. The following image shows examples of entering Polar and Cartesian Coordinates.

NOTE If no data is entered in the input fields and only points are picked, dimensions will NOT automatically be placed. You can manually place dimensions and constraints after the geometry is sketched.

FIGURE 2.13

Coordinate Types

While sketching you can right-click and change the coordinate types shown in the following image. You must select one option from the top and bottom of the menu:

Absolute Coordinate #: Values are relative to the 0,0 coordinates on the sketch.

Relative Coordinate @: Values are relative to the last selected point.

Polar Coordinate <: When defining a second or subsequent point, enter X value and angle.

Cartesian Coordinate >: Enter X and Y coordinates.

FIGURE 2.14

Polar Coordinates. By default you enter values using Polar coordinates by specifying the distance and angle of sketch elements. After entering the first value, press TAB to display a lock icon in the field, which constrains the cursor by the value that you entered and moves the cursor to the next field.

Cartesian Coordinates. The Rectangle and Point commands use Cartesian coordinates, where you specify X and Y coordinates. After entering the first value, press TAB to display a lock icon in the field, which constrains the cursor by the value that you entered and moves the cursor to the next field.

Persistent Dimensions

After entering values in the input fields, press ENTER to create the geometry, and automatically place dimension referred to as persistent dimensions. To disable persistent dimension and manually apply dimensions, turn off persistent dimensions by clicking Sketch tab > Constrain panel > click the drop arrow and click Persistent Dimensions as shown in the following image. Persistent dimensions are disabled until toggled back on. Persistent dimensions can also be turned on and off using the Heads-Up Display (HUD) settings on the Sketch tab in the Application Options dialog box. Also if you create geometry by clicking points, no dimensions will automatically be created.

FIGURE 2.15

Heads-Up Display Settings

Pointer Input and Dimension Input can be turned on/off; change to Cartesian or Polar Coordinates and change the Heads-Up Display (HUD) settings. Click the Sketch tab in the Application Options dialog box as shown in the following image on the left. Click the Setting button and the Heads-Up Display Settings dialog box will appear as shown in the following image on the right. Select the options in the dialog box to control the heads-up display options when dimensioning a sketch.

FIGURE 2.16

Pointer Input. When Pointer Input is on, the starting point for a sketch element is displayed as Cartesian coordinates (X and Y values) in a tooltip near the cursor. The default for second and subsequent points is relative Polar coordinates when defining lengths and angles.

When Pointer Input is off, the Cartesian coordinate displays the starting point of sketch geometry. Second and subsequent points are still displayed in the value input fields.

Dimension Input. When dimension input is on, the value input fields display polar coordinates when defining lengths and angles for a second point. The dimensional values change as you move the cursor. Press TAB to move to the next input field or click in another cell. After pressing the TAB key, the value will be locked and a lock icon will appear to the right of the value as shown in the following image. After a dimension's value is locked, the parametric dimension will be created after clicking a point or pressing the ENTER key. You can change the value in an input filed by either clicking in the field or pressing the TAB key until the field is highlighted and then typing in a new value.

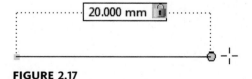

FIGURE 2.17

Line Command

The Line command, is one of the most powerful commands that you will use to sketch. Not only can you draw lines with it, but also you can draw an arc from the endpoint of a line segment. To start sketching lines, click the Line command from the Sketch tab > Draw panel as shown in the following image on the left, or right-click in a blank area in the graphics window and click Create Line from the marking menu as shown in the following image in the middle, or press the L key on the keyboard. After starting the Line command you will be prompted to click a first point, select a point in the graphics window, and then click a second point. The following image on the right shows the line being created with the dynamic input as well as the horizontal constraint.

FIGURE 2.18

You can continue drawing line segments, or you can sketch an arc from the endpoint. Move the cursor over the endpoint of a line segment or arc, and a small gray circle will appear at that endpoint as shown in the following image on the left. Click on the small circle, and with the left mouse button pressed down, move the cursor in the direction that you want the arc to go. Up to eight different arcs can be drawn, depending upon how you move the cursor. The arc will be tangent to the horizontal or vertical edges that are displayed from the selected endpoint. The following image on the right shows an arc that is normal to the sketched line being drawn.

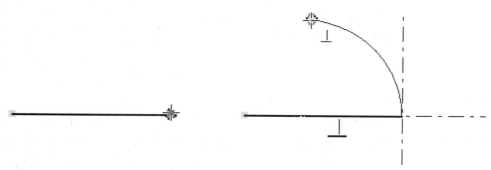

FIGURE 2.19

| When sketching, look at the bottom-right corner of the status bar (bottom of the screen) to see the coordinates, length, and angle of the objects that you are drawing. The following image shows the status bar when a line is being drawn. | **TIP** |

1.109 in, 0.330 in   x=0.605 in   y=0.000 in

FIGURE 2.20

Object Tracking – Inferred Points

If the Point Alignment On option is checked from the Sketch tab of the Application Options, as you sketch dashed lines will appear on the screen. These dashed lines represent the endpoints; midpoints; and theoretical intersections of lines, arcs, and center points of arcs and circles that represent their horizontal, vertical, or perpendicular positions. As the cursor gets close to these inferred points, it will snap to that location. If that is the point that you want, click that point; otherwise, continue to move the cursor until it reaches the desired location. When you select inferred points, no constraints (geometric rules such as horizontal, vertical, collinear, and so on) are applied from them. If the Point Alignment On option is unchecked, you can still infer points, move the cursor over a point, and move the cursor. Using inferred points helps create more accurate sketches. The following image shows the inferred points from two midpoints that represent their horizontal and vertical position.

FIGURE 2.21

Automatic Constraints

As you sketch, a small constraint symbol appears that represents geometric constraint(s) that will be applied to the object. If you do not want a constraint to be applied, uncheck the Constraint Persistent option under the Constrain panel drop arrow or hold down the CTRL key when you select the point. The following image shows a line being drawn from the arc, tangent to the arc, and parallel to the angled line, and the dynamic input is also displayed. The symbol appears near the object from which the constraint is coming. Constraints will be covered in the next section.

FIGURE 2.22

Scrubbing

As you sketch, you may prefer to apply a constraint different from the one that automatically appears on the screen. You may want a line to be perpendicular to a given line, for example, instead of being parallel to a different line. The technique to change the constraint is called *scrubbing*. To place a different constraint while sketching, move the cursor so it touches (scrubs) the other object to which the constraint should be related. Move the cursor back to its original location, and the constraint symbol changes to reflect the new constraint. The same constraint symbol will also appear near the scrubbed object, representing that it is the object to which the constraint is matched. Continue sketching as normal. The following image shows the top horizontal line being drawn with a perpendicular constraint that was scrubbed from the left vertical line. Without scrubbing the left vertical line, the applied constraint would have been parallel to the bottom line.

FIGURE 2.23

Common Sketch Tools

The following chart lists common 2D sketch commands that are not covered elsewhere in this chapter. Some commands are available by clicking the down arrow in the lower-right corner of the top command. Consult the help system for more information about these commands.

Tool	Function
Center-point Circle	Creates a circle by clicking a center point for the circle and then a point on the circumference of the circle.
Tangent Circle	Creates a circle that will be tangent to three lines or edges by clicking the lines or edges.
Three-Point Arc	Creates an arc by clicking a start and endpoint and then a point that will lie on the arc.
Tangent Arc	Creates an arc that is tangent to an existing line or arc by clicking the endpoint of a line or arc and then clicking a point for the other endpoint of the arc.
Center-Point Arc	Creates an arc by clicking a center point for the arc and then clicking a start and endpoint.
Two-Point Rectangle	Creates a rectangle by clicking a point and then clicking another point to define the opposite side of the rectangle. The edges of the rectangle will be horizontal and vertical.

(Continued)

Tool	Function
Three-Point Rectangle	Creates a rectangle by clicking two points that will define an edge and then clicking a point to define the third corner.
Fillet	Creates a fillet between two nonparallel lines, two arcs, or a line and an arc at a specified radius. If you select two parallel lines, a fillet is created between them without specifying a radius. When the first fillet is created, a dimension will be created. If many fillets are placed in the same operation, you choose to either apply or not apply an equal constraint.
Chamfer	Creates a chamfer between lines. There are three options to create a chamfer: both sides equal distances, two defined distances, or a distance and an angle.
Polygon	Creates an inscribed or a circumscribed polygon with the number of faces that you specify. The polygon's shape is maintained as dimensions are added.
Mirror	Mirrors the selected objects about a centerline. A symmetry constraint will be applied to the mirrored objects.
Rectangular Pattern	Creates a rectangular array of a sketch with a number of rows and columns that you specify.
Circular Pattern	Creates a circular array of a sketch with a number of copies and spacing that you specify.
Offset	Creates a duplicate of the selected objects that are a given distance away. By default, an equal-distance constraint is applied to the offset objects.
Trim	Trims the selected object to the next object it finds. Click near the end of the object that you want trimmed. While using the Trim command, hold down the SHIFT key to extend objects. If desired, hold down the CTRL key to select boundary objects.
Extend	Extends the selected object to the next object it finds. Click near the end of the object that you want extended. While using the Extend command, hold down the SHIFT key to trim objects. If desired, hold down the CTRL key to select boundary objects.

Selecting Objects

After sketching objects, you may need to move, rotate, or delete some or all of the objects. To edit an object, it must be part of a selection set. There are two methods that you can use to place objects into a selection set.

- You can select objects individually by clicking on them. To select multiple individual objects, hold down the CTRL or SHIFT key while clicking the objects. You can remove selected objects from a selection set by holding down the CTRL or SHIFT key and reselecting them. As you select objects, their color will change to show that they have been selected.
- You can select multiple objects by defining a selection window. Not all commands allow you to use the selection window technique and only allow single

selections. To define the window, click a starting point. With the left mouse button depressed, move the cursor to define the box. If you draw the window from left to right (solid lines), as shown in the following image on the left, only the objects that are fully enclosed in the window will be selected. If you draw the window from right to left (dashed lines), as shown in the following image on the right, all of the objects that are fully enclosed in the window *and* the objects that are touched by the window will be selected.

- You can use a combination of the methods to create a selection set.

When you select an object, its color will change according to the color style that you are using. To remove all of the objects from the selection set, click in a blank section of the graphics area.

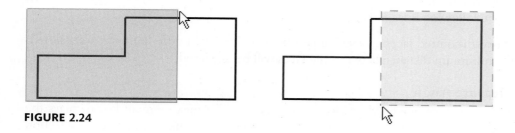

FIGURE 2.24

Deleting Objects

To delete objects first cancel the command that you are in by pressing the ESC key. Then select objects to delete, and then either press the DELETE key or right-click and select Delete from the menu as shown in the following image.

FIGURE 2.25

Measure Commands

Measure commands that assist in analyzing sketch, part, and assembly models are available. You can measure distances, angles, and loops, and you can perform area calculations. You can start the measure command first and then select the geometry, or select the geometry first and then start a measure command.

The Measure commands are on the Tools tab > Measure panel as shown in the following image. The following sections discuss these commands in greater detail.

FIGURE 2.26

Measure Distance

Measures the length of a line or edge, length of an arc, distance between points, radius and diameter of a circle, or the position of elements relative to the active coordinate system. A temporary line designating the measured distance appears, and the Measure Distance dialog box displays the measurement for the selected length, as shown in the following image.

If two disjointed faces of a single part or two faces from different parts are selected, the minimum distance between the faces will be displayed.

Measure Angle

Measures the angle between two lines or edges. The measurement box displays the angle based on the selection of two lines or edges or two lines defined by the selection of three points.

Measure Loop

Measures the length of closed or open loops defined by face boundaries or other geometry. When moving your cursor over a part face, all edges of the face will become highlighted. Clicking on this face will calculate the closed loop or perimeter of the shape.

Measure Area

Measures the area of enclosed regions or faces. Moving your cursor inside the closed outer shape will cause the outer shape and all holes (referred to as "islands") to also become highlighted. Clicking inside this shape will calculate the area of the shape.

When you click the arrow beside the measurement box, a menu similar to the following image will appear.

FIGURE 2.27

Brief explanations of each option follow:

Restart
Click to clear the measurement from the measurement box so that you can make another measurement.

Measure Angle
Click to change the measurement mode to Measure Angle.

Measure Loop
Click to change the measurement mode to Measure Loop.

Measure Area
Click to change the measurement mode to Measure Area.

Add to Accumulate
Click to add the measurement in the measurement box to accumulate a total measurement.

Clear Accumulate
Click to clear all measurements from the accumulated sum, resetting the sum to zero.

Display Accumulate
Click to display the sum of all measurements you have added to the accumulated sum.

Dual Unit
Click to display the measurement in another unit. The measurement in the second unit will be displayed below the first unit measurement.

Precision
Click to change the decimal display between showing all decimal places and showing the number of decimal places specified in the document settings for the active part or assembly.

Region Properties
Available by clicking the drop arrow on the bottom of the Measure panel. While in a sketch you can determine the properties such as the area, perimeter, and Moment of Inertia properties of a closed 2D sketch. All measurements are taken from the sketch coordinate system. The properties can be displayed in dual units, the default unit of the document, or a unit of your choice. The following image shows the region properties of two circles.

Autodesk Inventor 2012 School Edition

FIGURE 2.28

EXERCISE 2-1: CREATING A SKETCH WITH LINES

In this exercise, you create a new part file, and then create sketch geometry using basic construction techniques. In this exercise no dimensions will be created.

1. Click the New command, click the English tab, and then double-click *Standard (in).ipt,* or if inch is the default unit; from the left side of the Quick Access toolbar you can click the down arrow of the New icon, and select Part.
2. Click the Line command in the Draw panel.
3. Click on the origin point in the graphics window, move the cursor to the right approximately 4 inches, and, when the horizontal constraint symbol displays, click to specify a second point as shown in the following image. You may need to zoom back and pan the screen to draw the line.

FIGURE 2.29

NOTE Symbols indicate the geometric constraint. In the figure above, the symbol indicates that the line is horizontal. When you create the first entity in a sketch, make it close to final size. The length and angle of the line are displayed in the lower-right corner of the window to assist you.

4. Move the cursor up until the perpendicular constraint symbol displays beside the first line and then click to create a perpendicular line that is approximately 2 inches as shown in the following image on the left.
5. Move the cursor to the left and create a horizontal line approximately 1 inch, that is, parallel to the first horizontal line. The parallel constraint symbol is displayed as shown in the following image on the right.

FIGURE 2.30

6. Move the cursor down, and create a line that is parallel to the first vertical line and is approximately 1 inch.
7. Move the cursor left to create a horizontal line of approximately 2 inches.
8. Move the cursor up until the parallel constraint symbol is displayed, and a dotted alignment line appears as shown in the following image on the left. If the parallel constraint does not appear, move (scrub) the cursor over the inside vertical line to create a relationship to it. Click to locate the point.
9. Move the cursor to the left until the parallel constraint symbol is displayed, and a dotted alignment line appears as shown in the following image on the right.

FIGURE 2.31

   Then click to locate the point.
10. To close the profile, right-click and click Close from the menu.
11. Your screen should resemble the following image.
12. Right-click in the graphics screen, and click Done to end the Line command.
13. Right-click in the graphics screen again, and click Finish Sketch.
14. Close the file. Do not save changes. End of exercise.

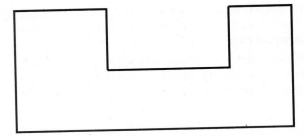

FIGURE 2.32

EXERCISE 2-2: CREATING A SKETCH WITH TANGENCIES

In this exercise, you create a new part file, and then you create a profile consisting of lines and tangent arcs.

1. Click the New command, click the English tab, and then double-click *Standard (inch).ipt*; or if inch is the default unit, from the left side of the Quick Access toolbar you can click the down arrow of the New icon, and select Part.
2. Click the Line command in the Draw panel.
3. Click on the projected origin point in the middle of the graphics window, and create a horizontal line to the right of the origin point and type 3 (inches will be assumed as the unit because the part file is based on the unit of inch) in the input field and then press ENTER.

> **NOTE** If the second point of the line lies off the screen, roll the mouse wheel away from you to zoom out, hold down the mouse wheel, and drag to pan the view.

4. Create a perpendicular line, move the cursor up, type 2 in the input field as shown in the following image on the left, and press ENTER.
5. In this step, you infer points, meaning that no sketch constraint is applied. Move the cursor to the intersection of the midpoints of the right-vertical line and bottom horizontal line. Dashed lines (inferred points) appear as shown in the following image on the right, and then click to create the line. No dimension was created since a value was not entered.

FIGURE 2.33

6. Next you create a line that is parallel to the bottom line. You scrub the bottom line by moving the cursor over the bottom line (do NOT click), and then move the cursor up and to the left until the vertical inferred line and the constraints are displayed as shown in the following image on the left, and then click to place the line.

7. Next you sketch an arc while in the line command. Click on the gray dot at the left end of the line, and hold and drag the endpoint to create a tangent arc. **Do not release the mouse button**.

8. Move the cursor over the endpoint of the first line segment until a coincident constraint (green circle) and the two tangency constraints at start and end points of the arc are displayed as shown in the following image on the right.

FIGURE 2.34

9. Release the mouse button to place the arc.
10. Right-click in the graphics window, and then click Done. Later in this chapter you will learn how to create dimensions.
11. Close the file. Do not save changes. End of exercise.

STEP 2—CONSTRAINING THE SKETCH

After you draw the sketch, you may want to add geometric constraints to it. Geometric constraints apply behavior to a specific object or create a relationship between two objects. An example of using a constraint is applying a vertical constraint to a line so that it will always be vertical. You could apply a parallel constraint between two lines to make them parallel to one another; then, as one of the line's angles changes, so will the other's. You can apply a tangent constraint to a line and an arc or to two arcs.

When you add a constraint, the number of constraints or dimensions that are required to fully constrain the sketch will decrease. On the bottom-right corner of Autodesk Inventor, the number of constraints or dimensions will be displayed similar to what is shown in the following image. A fully constrained sketch is a sketch whose objects cannot move or stretch.

FIGURE 2.35

To help you see which objects are constrained, Autodesk Inventor will change the color of constrained objects if a point on the sketch has a coincident constraint applied to the origin point or another point of an existing sketch. If you have not sketched a point coincident to the projected origin point or another point of an existing sketch, the color of the sketch will not change. You could apply a fix constraint instead of using the origin point, but it is not recommended. If you have not sketched a point coincident to the origin point or applied a fix constraint to the sketch, objects are free to move in their sketch plane.

NOTE Autodesk Inventor does not force you to fully constrain a sketch. However, it is recommended that you fully constrain a sketch, as this will allow you to better predict how a part will react when you change dimensions values.

Constraint Types

Autodesk Inventor has 12 geometric constraints that you can apply to a sketch. The following image shows the constraint types that can be applied from the marking menu and the symbols that represent them when the constrains are displayed in a sketch. Descriptions of the constraints follow.

FIGURE 2.36

Button	Tool	Function
⊥	Perpendicular	Lines will be repositioned at 90° angles to one another. The first line sketched will stay in its position, and the second will rotate until the angle between them is 90°.
//	Parallel	Lines will be repositioned so that they are parallel to one another. The first line sketched will stay in its position, and the second will move to become parallel to the first.
⌒	Tangent	An arc, circle, or line will become tangent to another arc or circle.
⌢	Smooth (G2)	A spline and another spline, line, or arc that connect at an endpoint with a coincident constraint will represent a smooth G2 (continuous curvature) condition.
⊥	Coincident	A point is constrained to lie on another point or curve (line, arc, etc.).
◎	Concentric	Arcs and/or circles will share the same center point.

(Continued)

Button	Tool	Function
⚡	Collinear	Two selected lines will line up along a single line; if the first line moves, so will the second. The two lines do not have to be touching.
=	Equal	If two arcs or circles are selected, they will have the same radius or diameter. If two lines are selected, they will become the same length. If one of the objects changes, so will the other object to which the Equal constraint command has been applied. If the Equal constraint command is applied after one of the arcs, circles, or lines has been dimensioned, the second arc, circle, or line will take on the size of the first one. If you select multiple similar objects (lines, arcs, etc.) before selecting this command, the constraint is applied to all of them. This rule applies to some of the other sketch constraints as well.
⎯	Horizontal	Lines are positioned parallel to the X axis, or a horizontal constraint can be applied between any two points in the sketch. The selected points will be aligned such that a line drawn between them will be parallel to the X axis.
∥	Vertical	Lines are positioned parallel to the Y axis, or a vertical constraint can be applied between any two points in the sketch. The selected points will be aligned such that a line drawn between them will be parallel to the Y axis.
🔒	Fix	Applying a fixed point or points will prevent the selected point from moving. The Fix constraint command overrides any other constraint. Any endpoint or segment of a line, arc, circle, spline segment, or ellipse can be fixed. Multiple points in a sketch can be fixed. If you select near the endpoint of an object, the endpoint will be locked from moving. If you select a line segment, the angle of the line will be fixed and only its length will be able to change. You can remove a fix constraint as needed. Deleting constraints will be covered later in this section.
⊔	Symmetry	Selected points defining the selected geometry are made symmetric about the selected line.

> **NOTE:** To fully constrain a base (first) sketch, constrain or dimension a point in the sketch to the center point or apply a fix constraint.

Adding Constraints

As stated previously in this chapter, you can apply constraints while you sketch objects. You can also apply additional constraints after the sketch is drawn. However, Autodesk Inventor will not allow you to overconstrain the sketch or add duplicate constraints. If you add a constraint that would conflict with another, you will be warned with the message, "Adding this constraint will overconstrain the sketch." For example, if you try to add a vertical constraint to a line that already has a horizontal constraint, you will be alerted. To apply a constraint, follow these steps:

1. Click a constraint from the Constrain panel, or right-click in the graphics window and click Create Constraint from the menu. Then click the specific constraint from the menu as shown in the previous image before the chart.
2. Click the object or objects to apply the constraints.

Showing and Deleting Constraints

To see the constraints that are applied to an object, use the Show Constraints command from the Constrain panel as shown on the left side of the following image. After issuing the Show Constraints command, select an object and a constraint icon and yellow squares that represent coincident constraints will appear with the constraints that are applied to the selected object—similar to what is shown in the middle of the following image. You can modify the size of the constraint toolbar by clicking Tools tab > Application Options, clicking on the Sketch tab, and modifying the size of the Constraint and DOF Symbol Scale. The following image on the right shows the scale increased from 1.0 to 1.5.

FIGURE 2.37

To show all the constraints for all of the objects in the sketch, do the following:

- While not in an operation, right-click in the graphics window, and from the menu, click Show All Constraints, or press the F8 key. To hide all the constraints, right-click in the graphics window, and click Hide All Constraints from the menu, or press the F9 key or right-click and click Hide All Constraints from the marking menu.
- If a constraint is added when sketch constraints are displayed, you will need to exit out of the active command and then press the F8 key again.
- When the constraints are shown, all the constraints in the sketch will appear. You can control which constraints are visible by right-clicking in the graphics window while not in a command and then clicking Constraint Visibility from the menu as shown in the following image on the left. The Constraint Visibility dialog will appear; uncheck the constraint type that you do not want to see. These settings apply only to the current sketch.

FIGURE 2.38

As you move the cursor over a constraint icon, the matching sketch constraint and the object that is linked to that constraint will be highlighted. The coincident constraint will appear as a yellow square at the point that the constraint exists. The perpendicular constraint will appear once at the location the constraint is applied. The following image on the left shows the cursor over the perpendicular constraint on the right vertical edge; the bottom horizontal and right vertical lines are highlighted. The following image on the right shows the cursor over the bottom horizontal line, and the constraints that are related to the line are highlighted.

FIGURE 2.39

To delete the constraint, either click on it and then right-click, or right-click while the cursor is over the constraint, or select Delete from the menu. You can also click on it, and press the DELETE key. The following image shows the parallel constraint being deleted.

To hide constraint icons, click the × on the right side of the constraint symbols toolbar.

FIGURE 2.40

CONSTRUCTION GEOMETRY

Construction geometry can help you create sketches that would be otherwise difficult to constrain. You can constrain and dimension construction geometry like normal geometry, but the construction geometry will not be recognized as a profile edge in the part when you turn the sketch into a feature. When you sketch, the sketches by default have a normal geometry style, meaning that the sketch geometry is visible in the feature. Construction geometry can reduce the number of constraints and dimensions required to constrain a sketch fully, and it can help to define the sketch.

For example, a construction circle that is tangent to the inside a hexagon (drawn with individual lines and not the Polygon command) can drive the size of the hexagon. Without construction geometry, the hexagon would require six constraints and dimensions. With construction geometry, it would require only three constraints and dimensions; the circle would have tangent or coincident constraints applied to it and the hexagon. You create construction geometry by changing the line style before or after you sketch geometry in one of the following two ways:

- After creating the sketch, select the geometry that you want to change and click the Construction icon on the Format panel as shown in following image.
- Before sketching, click the Construction icon on the Format panel, as shown in the following image on the left. All geometry created will be construction until Construction is unselected.

FIGURE 2.41

After turning the sketch into a feature, the construction geometry will be consumed with the sketch and is maintained in the sketch. When you edit a feature's sketch that you created with construction geometry, the construction geometry will reappear during editing and disappear when the part is updated. You can add or delete construction geometry to or from a sketch just like any geometry that has a normal style. In the graphics window, construction geometry will be displayed as a dashed line, lighter in color and thinner in width than normal geometry. The following image on the left shows a sketch with a construction line for the angled line. The angled line has a coincident constraint applied to it at every point that touches it. The image on the right shows the sketch after it has been extruded. Note that the construction line was not extruded.

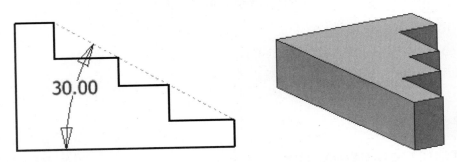

FIGURE 2.42

Snaps

Another method used to place geometry with a coincident constraint is snaps: midpoint, center, and intersection. After using a snap, a coincident constraint will be applied, and it will maintain the relationship that you define. For example, if you use a midpoint snap, the sketched point will always be in the middle of the selected object, even if the selected object's length changes.

To use snaps, follow these steps:

1. Start a sketch command, and right-click in the graphics window.
2. From the menu, select the desired snap.
3. Click on the object to which the sketched object will be constrained. For the intersection snap, select two objects.

FIGURE 2.43

Sketch Degrees of Freedom

While constraining and dimensioning a sketch there are multiple ways to determine the open degrees of freedom. When you add a constraint or dimension the number of constraints or dimensions that are required to fully constrain, the sketch will decrease.

Status Bar

On the bottom-right corner of Autodesk Inventor, the number of constraints or dimensions will be displayed similar to what is shown in the following image on the left. A fully constrained sketch is a sketch whose objects cannot move or stretch and the message Fully Constrained will appear in the bottom-right corner of the screen as shown in the following image on the right.

FIGURE 2.44

Degrees of Freedom

To see the areas in the sketch that are NOT constrained, you can display the degrees of freedom. While in a sketch right-click and click Show Degree of Freedom as shown in the following image in the middle. Lines and arcs with arrows will appear as shown in the following image on the right. As constraints and dimensions are added to the sketch, degrees of freedom will disappear. To remove the degree of freedom symbols from the screen, right-click and click Hide All Degrees of Freedom from the menu as shown in the following image on the right.

FIGURE 2.45

Dragging a Sketch

Another method to determine whether or not an object is constrained is to try to drag it to a new location. While not in a command, click a point or an edge, or select multiple objects on the sketch. With the left mouse button depressed, drag it to a new location. If the geometry stretches, it is underconstrained. For example, if you draw a rectangle that has two horizontal and two vertical constraints applied to it and you drag a point on one of the corners, the size of the rectangle will change, but the lines will maintain their horizontal and vertical behaviors. If dimensions are set on the object, they too will prevent the object from stretching.

EXERCISE 2-3: ADDING AND DISPLAYING CONSTRAINTS

In this exercise, you add geometric constraints to sketch geometry to control the shape of the sketch.

1. Click the New command, click the English tab, and double-click *Standard (in).ipt*.
2. Sketch the geometry as shown in the following image, with an approximate size of 2 inches in the X (horizontal) and 1.125 inches in the Y (vertical). Do not apply dimensions dynamically. Place the upper-left corner of the sketch on the projected origin point. Right-click in the graphics window, and then click Done. By starting the line at the origin point, that point is constrained to the origin.
3. Right-click in the graphics window, and click Show All Constraints, or press the F8 key. Your screen should resemble the following image.

FIGURE 2.46

4. If another constraint appears, place the cursor over it, right-click, and then click Delete from the menu.
5. On the Constrain panel, click the Parallel constraint icon.
6. Click the two angled lines. Depending upon the order in which you sketched the lines, the angles may be opposite of the following image on the left.
7. Press the ESC key twice to stop adding constraints.
8. Press the F8 key to refresh the visible constraints. Your screen should resemble the following image on the right.

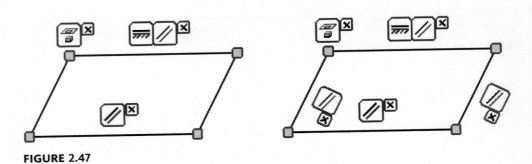

FIGURE 2.47

9. Click on the bottom horizontal line in the sketch and drag the line. Notice how the sketch changes its size, but not its general shape. Try to drag the top horizontal line. The line cannot be dragged as it is constrained.
10. Click the endpoint on the bottom-left horizontal line, and drag the endpoint. The lines remain parallel due to the parallel constraints.
11. Place the cursor over an icon for the parallel constraint for the angled lines, right-click, and click Delete from the menu as shown in the following image on the left. The parallel constraint that applies to both angled lines is deleted.
12. On the Constrain panel, click the Perpendicular constraint icon.
13. Click the bottom horizontal line and the angled line on the right side. Even though it may appear that the rectangle is fully constrained, the left vertical line is still unconstrained and can move.
14. Click the bottom horizontal line and the left angled line.
15. Press the F8 key to refresh the visible constraints. Your screen should resemble the following image on the right.
16. Right-click in the graphics window, and click Hide All Constraints, or press the F9 key.

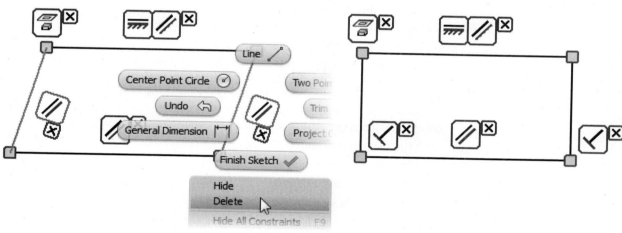

FIGURE 2.48

17. Drag the point at the lower-right corner of the sketch to verify that the rectangle can change size in both the horizontal and vertical directions, but its shape is maintained.
18. Close the file without saving the changes.
19. Click the New command, click the English tab, and then double-click *Standard (in).ipt*.
20. Sketch the geometry as shown with an approximate size of **2 inches** in the X and **1.375 inches** in the Y. Do not apply dimensions dynamically. Place the lower-left most point of the sketch on the projected center point. Right-click in the graphics window, and then click Done.

FIGURE 2.49

21. Inspect the constraints by dragging different points and edges.
22. Next you make the arcs equal in size. On the Constrain panel, click the Equal constraint icon or press the = key on the keyboard.
23. Click the arc on the left and the bottom arc.
24. Click the arc on the left and the arc on the right side.
25. Click the arc on the left and the arc on the top.
26. Next you align the line segments. On the Constrain panel, click the Collinear constraint icon or right-click in the graphics window, click Create Constraint, and then click Collinear. Note: In the following steps, if the constraint cannot be placed, you have a constraint that prevents it from being placed. Delete the constraint that is preventing the collinear constraint from being placed.
27. Click the two bottom horizontal lines.
28. Click the two top horizontal lines.
29. Click the two left vertical lines.
30. Click the two right vertical lines.
31. To stop applying the collinear constraint, either right-click and click Done from the marking menu or press the ESC key.
32. Next you will align the arcs. On the Constrain panel, click the Vertical constraint icon.
33. Click the center point of the bottom arc, and then click the center point of the top arc.
34. On the Constrain panel, click the Horizontal constraint icon.
35. Click the center point of the left arc, and then click the center point of the right arc.
36. To stop applying the constraints, right-click and click Done from the marking menu, or press the ESC key.
37. Display all of the constraints by pressing the F8 key. Your screen should resemble the following image.

FIGURE 2.50

38. Hide all of the constraints by pressing the F9 key.
39. Click on an endpoint in the sketch and drag the endpoint. Try dragging different points, and notice how the sketch changes.
40. Delete the geometry as shown in the following image on the right. Press the ESC key twice to cancel any command, click a point above and to the left of the top arc, drag a window so it encompasses the arc on the right, release the mouse button, and press the DELETE key on the key board.
41. Close the open line segments. Drag the open endpoints onto each other until your sketch resembles the following image on the right.

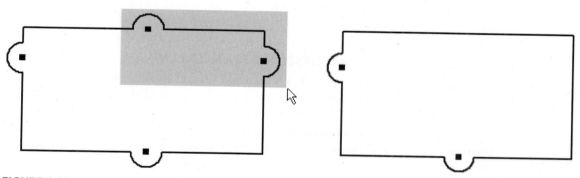

FIGURE 2.51

42. Next you center the arcs. On the Constrain panel, click the Vertical constraint icon.
43. Click the center point on the bottom arc and the midpoint of the top horizontal line as shown in the following image on the left.
44. On the Constrain panel, click the Horizontal constraint icon.
45. Click the center point on the left arc and the midpoint of the right vertical line as shown in the following image on the right.

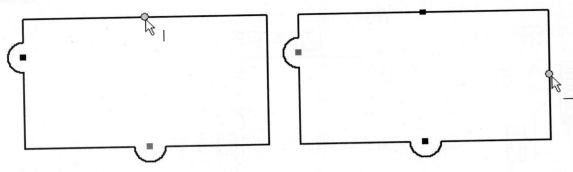

FIGURE 2.52

46. Click on different points and drag them, notice how the sketch changes shape, but the arcs are always centered as shown in the following image.

FIGURE 2.53

47. Close the file. Do not save changes. End of exercise. Note that dimensions would be added to completely define the sketch. Dimensions are covered in the next section.

STEP 3—ADDING DIMENSIONS MANUALLY

The last step to constraining a sketch is to add dimensions that were not added dynamically. The dimensions you place will control the size of the sketch and can also appear in the part drawing views when they are generated. When placing dimensions, try to avoid having extension lines go through the sketch, as this will require more cleanup when drawing views are generated. Click near the side from which you anticipate the dimensions will originate in the drawing views.

All dimensions that you create are parametric as well as the dynamic dimensions. Parametric means that they will change the size of the geometry. Parametric dimensions are manually created with either the General Dimension or Auto Dimension commands.

General Dimensioning

The General Dimension command can create linear, angle, radial, or diameter dimensions one at a time. The following image on the left shows an example of a dimensioned sketch. To start the General Dimension command, follow one of these techniques:

- Click the Create Dimension command from the Sketch tab > Constrain panel as shown in the following image on the right.

- Right-click in the graphics area and select Create Dimension from the marking menu.
- Press the shortcut key D.

FIGURE 2.54

When you place a linear dimension, the extension line of the dimension will snap automatically to the nearest endpoint of a selected line; when an arc or circle is selected, it will snap to its center point. To dimension to a quadrant of an arc or circle, see "Dimensioning to a Tangent of an Arc or Circle," later in this chapter.

After you select the General Dimension command, follow these steps to place a dimension:

1. Click a point or points to locate where the dimension is to start and end.
2. After selecting the point(s) to dimension, a preview image will appear attached to your cursor showing the type of dimension. If the dimension type is not what you want, right-click, and then select the correct style from the menu. After changing the dimension type, the dimension preview will change to reflect the new style.
3. Click a point on the screen to place the dimension.

The next sections cover how to dimension specific objects and how to create specific types of dimensioning with the Dimension command.

Dimensioning Lines

There are two techniques for dimensioning a line. Issue the Dimension command and do one of the following:

- Click near two endpoints, move the cursor until the dimension is in the correct location, and click.
- To dimension the length of a line, click anywhere on the line; the two endpoints will be selected automatically. Move the cursor until the dimension is in the correct location and click.
- To dimension between two parallel lines, click one line and then the next, and then click a point to locate the dimension.
- To create a dimension whose extension lines are perpendicular to the line being dimensioned, click the line and then right-click. Click Aligned from the menu, and then click a point to place the dimension.

Dimensioning Angles

To create an angular dimension, issue the General Dimension command, click on two lines whose angle you want to define, move the cursor until the dimension is in the correct location, and place the dimension by clicking on a point.

Dimensioning Arcs and Circles

To dimension an arc or circle, issue the General Dimension command, click on the circle's circumference, move the cursor until the dimension is in the correct location, and click. To dimension the angle of an arc, click on the arc; click the arc center point or click an endpoint of the arc, the center point, and then the other endpoint; and then locate the dimension. By default, when you dimension an arc, the result is a radius dimension. When you dimension a circle, the default is a diameter dimension. To change the radial dimension to diameter or a diameter to radial, right-click before you place the dimension and select the other style from the menu.

You can dimension the angle of the arc. Starting the Dimension command, click on the arc's circumference, click the center point of the arc, and then place the dimension or click the center point and then click the circumference of the arc.

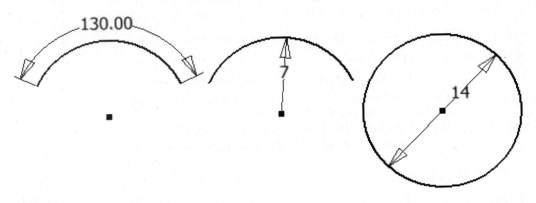

FIGURE 2.55

Dimensioning to a Tangent of an Arc or Circle

To dimension to a tangent of an arc or circle, follow these steps:

1. Issue the General Dimension command.
2. Click a line that is parallel to the tangent arc or circle that will be dimensioned.
3. Place the cursor over the tangent arc or circle that should be dimensioned.
4. Move the cursor over the tangent until the constraint symbol changes to reflect a tangent, as shown on the left side of the following image.
5. Click to accept the dimension type, and then move the cursor until the dimension is in the correct location. Click as shown on the right side of the following image.

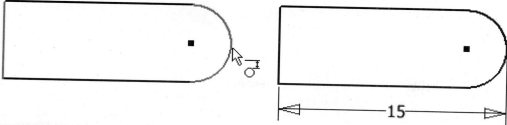

FIGURE 2.56

To dimension to two tangents, follow these steps:

1. Issue the General Dimension command.
2. Click an arc or circle that includes one of the tangents to which it will be dimensioned.
3. Place the cursor over the tangent edge of the second arc or circle to which it will be dimensioned.
4. Move the cursor over the area that is tangent until the constraint symbol appears, as shown on the left in the following image.
5. Click and then move the cursor until the dimension is in the correct location, and then click as shown in the following image on the right.

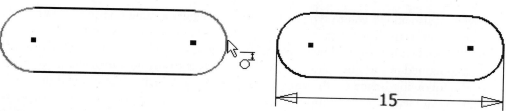

FIGURE 2.57

Entering and Editing a Dimension Value

After placing the dimension, you can change its value. Depending on your setting for editing dimensions when you created them, the Edit Dimension dialog box may or may not appear automatically after you place the dimension. To set the Edit Dimension option, do one of the following:

1. Click the Tools tab > Options panel > Application Options. On the Sketch tab of the Options dialog box, select Edit dimension when created as shown on the left side of the following image.
2. Or set this option by right-clicking in the graphics area while placing a dimension and selecting Edit Dimension from the menu as shown in the following image on the right. This method will change the Application Option sketch tab Edit Dimension.

If the Edit dimension when created option is checked, the Edit Dimension dialog box will appear automatically after you place the dimension. Otherwise, the dimension will be placed with the default value, and you will not be prompted to enter a different value.

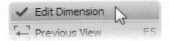

FIGURE 2.58

To edit a dimension that has already been created, double-click on the value of the dimension, and the Edit Dimension dialog box will appear, as shown in the following image. Enter the new value and unit for the dimension; then either press ENTER or click the checkmark in the Edit Dimension dialog box. If no unit is entered, the units that the file was created with will be used. When inputting values, enter the exact value—do not round up or down. The accuracy of the dimension that is displayed in a sketch is set in the Document Setting. For example, if you want to enter 4 1/16 decimally enter 4.0625 not 4.06.

FIGURE 2.59

Fractions

Inventor also allows you to type in a fraction anywhere a value is required. When the length unit in the Tools tab > Options > Document Settings is set to a metric unit the decimal equivalent will be displayed in the graphics window but the fraction will be maintained in the edit dialog box. If the unit is set to any non-metric unit as shown in the following image on the left and a fraction is entered, a fraction will display in the graphics window. After inputting a fraction you can click on the right-faced arrow and set the type of dimension to display Decimal, Fractional, or Architectural as shown in the following middle image. For example, enter 4 1/16, not 4-1/16. Inventor would interpret the - as part of an equation and would return the value 3.9375. The following image on the right shows the fraction displayed in the graphics window.

FIGURE 2.60

> When placing dimensions, it is recommended that you place the smallest dimensions first. This will help prevent the geometry from flipping in the opposite direction.

NOTE

Repositioning a Dimension

Once you place a dimension, you can reposition it, but the origin points cannot be moved. Follow these steps to reposition a dimension:

1. Exit the current operation either by pressing ESC twice or right-clicking and then selecting Done from the menu.
2. Move the cursor over the dimension until the move symbol appears as shown in the following image.
3. With the left mouse button depressed, move the dimension or value to a new location and release the button.

FIGURE 2.61

Overconstrained Sketches

As explained in the "Adding Constraints" section, Autodesk Inventor will not allow you to overconstrain a sketch or add duplicate constraints. The same is true when adding dimensions. If you add a dimension that will conflict with another constraint or dimension, you will be warned that this dimension will overconstrain the sketch or that it already exists. You will then have the option to either not place the dimension or place it as a driven dimension.

A driven dimension is a reference dimension. It is not a parametric dimension—it reflects the size of the points to which it is dimensioned, if the part changes the driven dimension will be updated to reflect the new value. A driven dimension will appear with parentheses around the dimensions value—for example, (30). When you place a dimension that will overconstrain a sketch, a dialog box will appear similar to the one in the following image. You can either cancel the operation and no dimension will be placed or accept the warning and a driven dimension will be created.

FIGURE 2.62

Autodesk Inventor gives you an option for handling overconstrained dimensions. To set the overconstrained dimensions option, click Tools tab > Options panel > Application Options, and on the Sketch tab of the Options dialog box, change the Overconstrained dimensions option as shown in the following image. You have the following two options:

- Apply driven dimension

When checked, this option automatically creates a driven dimension without warning you of the condition.

- Warn of overconstrained condition

When checked, this option causes a dialog box to appear stating that the dimension will overconstrain the sketch. Click Cancel not to place a driven dimension, or click Accept to place a driven dimension.

FIGURE 2.63

Another option for controlling the type of dimension that you create is to use the Driven Dimension command on the Format panel. If you issue the Driven Dimension command, any dimension you create will be a driven dimension. If you do not issue the command, you create a regular dimension. The following image shows the Driven Dimension command on the Format panel in its normal condition. The same Driven Dimension command can be used to change an existing dimension to either a normal or driven dimension by selecting the dimension and clicking the Driven Dimension command. Driven dimensions are represented in the sketch with parenthesis around the text as shown in the following image on the right, whereas parametric dimensions do not have parenthesis around the text.

| TIP | Avoid over using driven dimensions as they do not parametrically control the size of the profile, and reference dimensions are used only in the manufacturing process to verify the size of the part. |

FIGURE 2.64

EXERCISE 2-4: DIMENSIONING A SKETCH

In this exercise, you add dimensional constraints to a sketch. Note: this exercise assumes that the "Edit dimension when created" and "Autoproject part origin on sketch create" options are checked in the Options dialog box under the Sketch tab. Experiment with Autodesk Inventor's color schemes to see how the sketch objects changes color to represent if they are constrained.

1. Click the New command and click the English tab, and then double-click *Standard (in).ipt*.
2. Start sketching by starting the line command and click the origin point, move the cursor to the right, type **5** in the distance input field, press the TAB key, and then click a point when the horizontal constraint is previewed above the input field for degrees as shown in the following image on the left.
3. Next you place an angle line and a dynamic dimension to define the angle. Press the TAB key and type **150** for the angle input field, press the TAB key, and then click a point to the upper right as shown in the following image on the right. The distance should be about 2 inches but the dimension is not needed to define this sketch.

FIGURE 2.65

4. Sketch the geometry as shown in the following image. When sketching, ensure that a perpendicular constraint is not applied between the two angled lines. Hold down the CTRL key while sketching the top angle line to prevent sketch constraints from being applied. The arc should be tangent to both adjacent lines.

FIGURE 2.66

5. Add a horizontal constraint between the midpoint of the left vertical line and the center of the arc as shown in the following image on the left.
6. Add a vertical constraint between the endpoints of the angled lines nearest the right side of the sketch as shown in the following image on the right.

FIGURE 2.67

7. Add an equal constraint between the two angled lines by pressing the = key and then select the two angled lines.
8. Click the General Dimension command in the Constrain panel.
9. Add a radial dimension by clicking the arc. Drag the dimension to the left, click a point to locate it, enter **1.5** (if the Edit Dimension dialog box did not appear, double-click on the dimension and change the dimension to 1.5), and click the checkmark in the Edit Dimension dialog box.

NOTE To set the Edit Dimension option to appear when placing sketch dimensions, right-click in the graphics area while placing a dimension and select Edit Dimension from the menu.

10. Add a vertical dimension by clicking the vertical line. Drag the dimension to the left, click a point to locate it, enter **5**, and click the checkmark. When complete, your sketch should resemble the following image on the left. Notice on the bottom right of the status bar that the sketch requires 1 dimension as shown in the following image on the right.

FIGURE 2.68

11. Add an overall horizontal dimension by clicking the vertical line (not an endpoint). Move the cursor near the right tangent point of the arc until the glyph of dimension to a circle appears, as shown in the following image on the left. Click, drag the dimension down, click a point to locate it, enter **12**, and click the checkmark. When complete, your sketch should be fully constrained and resemble the following image on the right.

FIGURE 2.69

12. Edit the value of the dimensions, and examine how the sketch changes. The arc should always be in the middle of the vertical line. All of the dimensions are parametric, even the dimensions that were placed dynamically while sketching.
13. Delete the horizontal constraint between the center of the arc and the midpoint of the vertical line.
14. Delete the vertical constraint between the two angled lines.
15. Practice adding other types of constraints and dimensions.
16. Close the file. Do not save changes. End of exercise.

APPLYING YOUR SKILLS

Skill Exercise 2-1

In this exercise, you create a sketch and then add geometric and dimensional constraints to control the size and shape of the sketch. Start a new part file based on the *Standard (in).ipt*, and create the fully constrained sketch as shown in the following image. Assume that the top and bottom horizontal lines are collinear, the center points of the arcs are aligned vertically, and the sketch is symmetric about the left and right sides. One of the bottom angled lines should be coincident with the center point of the lower arc (if the arc is drawn via the line command, the center point of the arc will automatically be coincident with the line it was drawn from). When done, close the file and do not save the changes.

FIGURE 2.70

Skill Exercise 2-2

In this exercise, you create a sketch with linear and arc shapes, and then add geometric and dimensional constraints to fully constrain the sketch. Start a new part file based on the *Standard (in).ipt*, and create the fully constrained sketch as shown in the following image. First create the two outer circles and align their center points horizontally. Then create the two lines, trim the two circles, and place a vertical constraint between the line endpoints on both ends. The dimension on the arcs define the arc's radius. When done, close the file and do not save the changes.

FIGURE 2.71

CHECKING YOUR SKILLS

Use these questions to test your knowledge of the material in this chapter.

1. True ___ False ___ When you sketch, constraints are not applied to the sketch by default.
2. True ___ False ___ When you sketch and a point is inferred, a constraint is applied to represent that relationship.
3. True ___ False ___ A sketch does not need to be fully constrained.
4. True ___ False ___ When working on an mm part, you cannot use English (inch) units.
5. True ___ False ___ After a sketch is constrained fully, you cannot change a dimension's value.
6. True ___ False ___ A driven dimension is another name for a parametric dimension.
7. True ___ False ___ Dimensions placed dynamically are not parametric.
8. True ___ False ___ You can import only 2D AutoCAD data into Autodesk Inventor.
9. Explain how to draw an arc while in the Line command.
10. Explain how to remove a geometric constraint from a sketch.
11. Explain how to change a vertical dimension to an aligned dimension while you create it.
12. Explain how to create a dimension between two quadrants of two arcs.
13. True ___ False ___ AutoCAD needs to be installed to insert AutoCAD geometry.
14. True ___ False ___ When a sketch is extruded that contains construction geometry, the construction geometry is deleted.
15. Explain how to change the unit type in a part file.

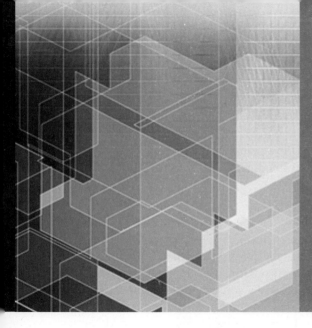

CHAPTER 3

Creating and Editing Sketched Features

INTRODUCTION

After you have drawn, constrained, and dimensioned a sketch, your next step is to turn the sketch into a 3D part. This chapter takes you through the process to create and edit sketched features.

OBJECTIVES

After completing this chapter, you will be able to perform the following:

- Understand what a feature is
- Understand Autodesk Inventor's browser
- Use direct manipulation techniques to create and edit a part
- Extrude a sketch into a part
- Revolve a sketch into a part
- Edit features of a part
- Edit the sketch of a feature
- Make an active sketch on a plane
- Create sketched features using one of three operations: cut, join, or intersect
- Project edges of a part

UNDERSTANDING FEATURES

After creating, constraining, and dimensioning a sketch, the next step in creating a model is to turn the sketch into a 3D feature. The first sketch of a part that is used to create a 3D feature is referred to as the base feature. In addition to the base feature, you can create sketched features in which you draw a sketch on a planar face or work plane, and you can either add or subtract material to or from existing features in a part. Use the Extrude, Revolve, Sweep, or Loft commands to create sketched features in a part. You can also create placed features such as fillets, chamfers, and holes by applying them to features that have been created. Placed features will be covered in Chapter 4. Features are the building blocks in creating parts.

A plate with a hole in it, for example, would have a base feature representing the plate and a hole feature representing the hole. As features are added to the part, they appear in the browser, and the history of the part or assembly, that is, the order in which the features are created or the parts are assembled, is shown. Features can be edited, deleted from, or reordered in the part as required.

Consumed and Unconsumed Sketches

You can use any sketch as a profile in feature creation. A sketch that has not yet been used in a feature is called an unconsumed sketch. When you turn a 2D sketched profile into a 3D feature, the feature consumes the sketch. The following image shows an unconsumed sketch in the browser on the left and a consumed sketch in the browser on the right. If desired, you can display more information in the browser about the features by clicking on the filter icon in the browser and click Show extended Names.

FIGURE 3.1

Although a consumed sketch is not visible as you view the 3D feature, you may need to access sketches and change their geometric or dimensional constraints in order to modify their associated features. A consumed sketch can be accessed from the browser by right-clicking and selecting Edit Sketch from the menu or by using the direct-manipulation mini-toolbar. The editing process will be covered later in this chapter. You may also access the sketch by starting the Sketch command and selecting the sketch from the browser. The following image on the left shows the unconsumed sketch, and the image on the right shows the extruded solid that consumes the sketch.

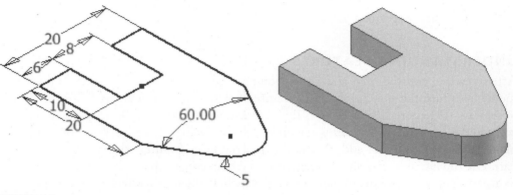

FIGURE 3.2

UNDERSTANDING THE BROWSER

The Autodesk Inventor browser, by default, is docked along the left side of the screen and displays the history of the file. In the browser you can create, edit, rename, copy, delete, and reorder features or parts. You can expand or collapse the browser to display the history of the features (the order in which the features were created) by clicking the + and − on the left side of the part or feature name in the browser. An alternate method of expanding the browser is to place your cursor on top of a feature icon but not click. The item in the browser automatically expands. To expand all the features, move the cursor into a blank area in the browser, right-click, and click Expand All from the menu.

The following image shows a browser with features of the part expanded.

FIGURE 3.3

As parts grow in complexity, so will the information found in the browser. Dependent features are indented to show that they relate to the item listed above it. This is referred to as a *parent-child relationship*. If a hole is created in an extruded rectangle, for example, and the extrusion is deleted, the hole will also be deleted.

Each feature in the browser is given a default name. The first extrusion, for example, will be named Extrusion1, and the number in the name will sequence as you add similar features. The browser can also help you to locate features in the graphics area. To highlight a feature in the graphics window, simply move your cursor over the feature name in the browser.

To zoom in on a selected feature, right-click on the feature's name in the browser and select Find in Window on the menu or press the END key on your keyboard. The browser itself functions similarly to a toolbar except that you can resize it while it is docked. To close the browser, click the X in its upper-right corner. If the browser is not visible on the screen, you can display it by clicking the View tab > Windows panel > User Interface > Browser as shown in the following image.

FIGURE 3.4

Specific functionality of the browser will be covered throughout this book in the pertinent sections. A basic rule is to either right-click or double-click the feature's name or icon to edit or perform a function on the feature.

SWITCHING ENVIRONMENTS

Up to this point, you have been working in the sketch environment where the work is done in 2D. The next step is to turn the sketch into a feature. To do so, you need to exit the *sketch environment* and enter the *part environment*. A number of methods can be used to accomplish this transition:

- Click the Finish Sketch command on the right side of the Sketch tab > Exit panel as shown in the following image on the left.
- In a blank area in the graphics window right-click and click Finish 2D Sketch from the menu as shown in the middle of the following image.
- Press the S key on the keyboard.
- Click the Model tab as shown in the following image on the right and click on a command.
- You can also right-click in the graphics area, select Create Feature from the menu, and then click the command you need to create the feature. Only the commands that are applicable to the current situation will be available.
- Enter a shortcut key to initiate one of the feature commands.

FIGURE 3.5

MODEL COMMANDS

When you exit the sketch environment, the Model Ribbon is active. The following image shows the Model tab. Many of these commands will be covered throughout this book.

FIGURE 3.6

DIRECT MANIPULATION

Direct manipulation allows you to start common commands and operations by clicking directly on the geometry in the graphics window. The direct manipulation tools appear as mini-toolbars or in-canvas buttons. Each direct manipulation option is explained next.

Mini-Toolbars

While not in a command you have different options depending upon what geometry is selected. After selecting the geometry click on the desired option. The different options will be explained throughout the book as they pertain to the topic.

Sketch

While not in a command, not in the sketch environment, and you select a sketch, the mini-toolbar appears and provides you the option of performing Extrude, Revolve, Hole, or Edit Sketch operations.

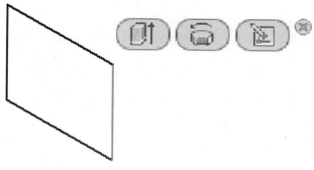

FIGURE 3.7

Face

While not in a command and you select a face, a mini-toolbar appears that allows you to quickly create a new sketch, edit an existing sketch, or edit a feature.

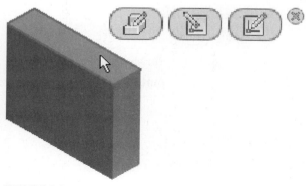

FIGURE 3.8

Edge

While not in a command and you select an edge, the mini-toolbar appears and provides you the option of performing a fillet or chamfer operation on the selected edge.

FIGURE 3.9

Mini-Toolbar–Command Options

After starting selected commands, the mini-toolbar will display command options in the graphics screen next to a selected object. The mini-toolbar options are presented via buttons and selection tags. The in-canvas buttons and selection tags display command options and prompts such as profile, face, and axis. Min-toolbar display is enabled for Extrude, Revolve, Fillet, Chamfer, Hole and Face Draft commands. The following image on the left shows the mini-toolbar buttons and the selection tag (profile) when extruding the first profile. You can move the mini-toolbar by clicking and dragging the vertical obround button in the upper left corner of the mini-toolbar. When the Pin Mini-Toolbar Position option is unchecked the mini-toolbars position will move as the arrow on the face is dragged. When checked the mini-toolbar will stay in its current position when the arrow on the face is dragged. The Fade option in the mini-toolbar will fade when the cursor moves away from the mini-toolbar. The option to pin and fade the mini-toolbar is controlled by clicking on the bottom-right button on the mini-toolbar as shown in the following image on the right. The mini-toolbar display options are covered throughout the book where the command is covered.

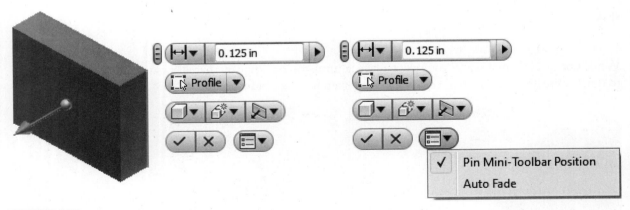

FIGURE 3.10

EXTRUDE A SKETCH

The most common method for creating a feature is to extrude a sketch and give it depth along the Z axis. Before extruding, it is helpful to view the part in an isometric view. When you exit the first sketch the viewpoint will automatically change to the home view. Autodesk Inventor previews the extrusion depth and direction in the graphics window.

To extrude a sketch, click the Extrude command on the Create panel bar as shown in the following image on the left, or while not in a sketch click on the sketch and the mini-toolbar will appear, click the Create Extrude button as shown in the following image in the middle, or press the shortcut key E. Alternately, you can right-click in the graphics screen and select Extrude on the marking menu as shown in the following image on the right.

FIGURE 3.11

After starting the Extrude command, the in-canvas display options will appear in the graphics window and the Extrude dialog box will appear. The following image on the left shows the in-canvas extrude options that are available when extruding the first sketch and the image on the right shows the operation button that is available when creating additional extrude features. You can either enter data in the dialog box or via the in-canvas display. For the mini-toolbar buttons, click the down arrow next to each button to see more options and click and drag the arrow to change the distance or taper. The mini-toolbar buttons will change depending upon what options are selected. If selection tags have a red arrow that command option needs to be satisfied before the command can be completed.

FIGURE 3.12

After you start the command, the Extrude dialog box also appears as shown in the following image. The Extrude dialog box has two tabs: Shape and More. When you make changes in the dialog box, the shape of the sketch will change in the graphics area to represent these values and options. When you have entered the values and the options you need, click the OK button to create the extruded feature. The extrude options will be explained in the next sections.

FIGURE 3.13

> **NOTE** When an arrow is red in a dialog box or in a selection tag in the mini-toolbar area that prompts for profiles, faces, and axes, the color indicates that Autodesk Inventor needs input from you to complete that function.

Shape

The Shape tab gives you options to specify the profile to use, operation, extents, and output type. The options are described below.

	Profile	Click this button to choose the sketch that you wish to extrude. If there are multiple closed profiles, you will need to select which sketch area you want to extrude. If there is only one possible profile, Autodesk Inventor will select it for you, and you can skip this step. If you select the wrong profile or sketch area, click the Profile button again and choose the desired profile or sketch area. To remove a selected profile, hold the CTRL key and click the area you wish to remove.
	Solid	If there are multiple solid bodies, click this button to choose the solid body(ies) to participate in the operation.

Operation

This is the unlabeled middle column of buttons. If this is the first sketch that you create a solid from, it is referred to as a base feature, and only the top button is available. The operation defaults to Join, which is the top button. Once the base feature has been established, you can extrude a sketch, adding or removing material from the part by using the Join or Cut options, or you can retain the common volume between the existing part and the newly defined extrude operation using the Intersect option.

	Join	Adds material to the part.
	Cut	Removes material from the part.
	Intersect	Removes material, keeping what is common to the existing part feature(s) and the new feature.
	New Solid	Creates a new solid body. The first solid feature created uses this option by default. Select to create a new body in a part file with an existing solid body.

Extents

The Extents option determines how the extruded sketch will be terminated. There are five options from which to choose as shown in the following image, but like the operation section, some options are not available until a base feature exists.

FIGURE 3.14

Distance

This option determines that the sketch will be extruded a specified distance.

To Next

This option determines that the sketch will be extruded until it reaches a plane or face. The sketch must be fully enclosed in the area to which it is projecting; if it is not fully enclosed, use the To termination with the Extend to Surface option. Click the Direction button to determine the extrusion direction.

To

This option determines that the sketch will be extruded until it reaches a selected face or plane. To select a point (midpoint or endpoint), plane, or face to end the extrusion, click the Select Surface to end the feature creation button, as shown in the following image, and then click a face or plane at which the extrusion should terminate.

FIGURE 3.15

Between

This option determines that the extrusion will start at a selected plane or face and stop at another plane or face. Click the Select surface to start feature creation button as shown in the following image on the left, and then click the face or plane where the extrusion will start. Then click the Select surface to end the feature creation button, as shown in the following image on the right and then click the face or plane where the extrusion will terminate.

FIGURE 3.16

All

This option determines that the sketch will be extruded all the way through the part in one or both directions.

Distance Value

If the extent is set to Distance, you can either enter a value in the dialog box or in the in-canvas display where the sketch will be extruded or click and drag the arrow in the in-canvas area. Another option is to click the arrow to the right where the distance value is displayed, from here you have an option to click two points to determine a value, display the dimensions of previously created features to select from, or select from the list of the recent values used. After a value is determined, a preview image appears in the graphics area to show how the extrusion will look.

A preview image appears in the graphics area, and the corresponding value appears in the distance area in the dialog box and in the in-canvas area.

If values and units appear in red when you enter them, the defined distance is incorrect and should be corrected. For example, if you entered too many decimal places (e.g., 2.12.5) or an incorrect unit for the dimension value, the value will appear in red. You will need to correct the error before the extrusion can be created.

Direction

There are four buttons from which to choose for determining the direction. Choose from the first two to flip the extrusion direction, click the third button (symmetric) to have the extrusion go equal distances in the negative and positive directions, and click the asymmetric option to define different distances for the negative and positive direction. With the midplane option if the extrusion distance is 2 inches, for example, the extrusion will go 1 inch in both the negative and positive Z directions.

FIGURE 3.17

Output

Two options are available to define the type of output that the Extrude command will generate:

| | Solid | Extrudes the sketch, and the result is a solid body. |
| | Surface | Extrudes the sketch, and the result is a surface. |

Match Shape

You can use the match shape option when working with an open profile that you want to extrude. The edges of the open profile are extended until they intersect geometry of the model. This provides a "flood-fill"–type effect for the extrude feature.

When you convert a sketch into a part or feature, the dimensions on the sketch are consumed (they disappear). When you edit the feature, the dimensions reappear. The dimensions can also be displayed when drawing views are made. For more information on editing parts or features, see the "Editing 3D Parts" section later in this chapter.

More

The More tab, as shown in the following image, contains additional options to refine the feature being created.

Alternate Solution
Alternate Solution terminates the feature on the most distant solution for the selected surface. An example is shown in the following image.

Minimum Solution
Minimum Solution option when checked terminates the feature on the first possible solution for the selected surface. An example is shown in the following image.

FIGURE 3.18

Taper
Taper extrudes the sketch and applies a taper angle to the feature. To extend the taper angle out from the part, give the taper angle a positive or negative number. This increases or decreases the volume of the resulting extruded feature. You can also click on the sphere in the mini-toolbar area and an arrow will appear as shown in the following image. Drag the arrow or enter a value.

FIGURE 3.19

Infer iMates
Check this box to automatically create an iMate on a full circular edge. Autodesk Inventor attempts to place the iMate on the closed loop most likely to be used.

EXERCISE 3-1: EXTRUDING A SKETCH

In this exercise, you will create a base feature by extruding an existing profile. You will examine the direction options available in the Extrude dialog box.

1. Open *ESS_E03_01.ipt* from the Chapter 03 subfolder.
2. From the Model tab > Create panel, click the Extrude command or click on the sketch and in the mini-toolbar click the Create Extrude button. Since there is only one closed profile, the profile is automatically selected.
3. In the Extrude dialog box, set the Distance to **.5 in**.
4. In the graphics window, select the arrow on the extrusion in the graphics window and drag the arrow until a distance of **.75 in** is displayed in the value field of the Extrude dialog box and in mini-toolbar display as shown in the following image, and then release the mouse button.

FIGURE 3.20

5. Click the More tab in the Extrude dialog box or click the sphere at the start of the arrow in the graphics window.
6. Adjust the value of the Taper to **20**, by entering 20 in the Extrude dialog box. In the graphics window, you could also have applied taper by clicking on the sphere on the back of the extrusion and drag the arrow.

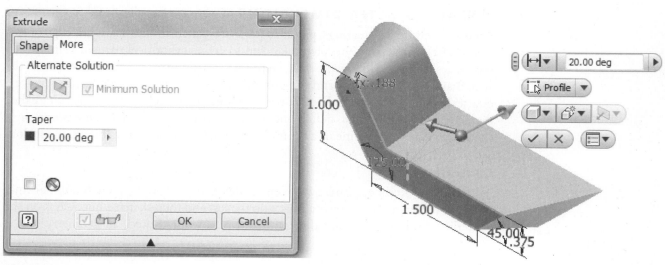

FIGURE 3.21

7. In the graphics window, click and drag the active arrow (for the taper) to adjust the taper to 15° and then change it back to 10°.
8. Click the Shape tab in the Extrude dialog box.
9. Change the direction of the extrusion to go in the negative Y direction by selecting the Direction 2 button (second from the left button) on the direction area in the dialog box. This will flip the direction of the extrusion.
10. In the dialog box, change the direction to symmetric (go evenly in both directions) by selecting the second from the right button.
11. In the mini-toolbar display area, change the extent to asymmetric as shown in the following image on the left.
12. In the graphics window, click and drag the arrows to different values as shown in the following image on the right.

FIGURE 3.22

13. Practice changing the values and directions to see the results. When done, click OK or click the green check mark in the mini-toolbar display area and the extrusion has been created. Later in this chapter you will learn how to edit features.
14. Close the file. Do not save changes. End of exercise.

LINEAR DIAMETER DIMENSIONS

Another method for creating a part is to revolve a sketch around a straight edge or axis (centerline). You can use revolved sketches to create cylindrical parts or features. To revolve a sketch, you follow the same steps that you did to extrude a sketch. Create the sketch and add constraints and dimensions. When revolving a sketch, you revolve a quarter section of the completed part. To define the sketch, you can place radial or diameter dimensions that define the sketch. To place diameter dimensions on a sketch, you use a centerline or a normal line to define the linear diameter dimension.

Centerline

To create a centerline, activate the Centerline command on the Sketch tab > Format panel, as shown in the following image on the left and then sketch a line. Click on the centerline to deactivate it. If the command is not deactivated, all geometry that is sketched will be a centerline. Or you can change an existing sketch entity to a centerline by selecting an existing sketched entity and then select the Centerline command. To create a linear diameter dimension by using a centerline, follow these steps:

1. Draw a sketch that represents a quarter section of the finished part.
2. Draw a centerline or change an existing line into a centerline that the sketch will be revolved around.
3. Start the Dimension command.
4. Click the centerline, or a point or edge to be dimensioned.
5. Click the centerline, or a point or edge to be dimensioned. It does not matter the order that the entities are selected; however, one of the selections needs to be the centerline not just an endpoint of the centerline.
6. Move the cursor until the diameter dimension is in the correct location and click.
7. The following image on the right shows the linear diameter dimension created using the centerline to define the axis of revolution.

FIGURE 3.23

Normal Line

When a normal line will be used to revolve the sketch around, you can create linear diameter (diametric) dimensions for sketches that represent a quarter outline of a revolved part. To create a linear diameter dimension, follow these steps:

1. Draw a sketch that represents a quarter section of the finished part.
2. Draw a line, if needed, or use a line in the sketch that the sketch will be revolved around.
3. Start the Dimension command.
4. Click the line (not an endpoint) that will be the axis of rotation.
5. Click the other point or line to be dimensioned.
6. Right-click and select Linear Diameter from the menu as shown in the following image on the left.
7. Move the cursor until the diameter dimension is in the correct location and click. The following image on the right shows a sketch that represents a quarter section of a part with a linear diameter dimension placed.

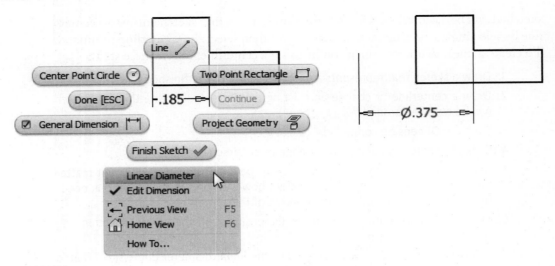

FIGURE 3.24

REVOLVE A SKETCH

After defining the sketch that will be revolved, click the Revolve command on the Create panel as shown in the following image on the left; or while not in a sketch, in the graphics window click on the sketch geometry and the mini-toolbar will appear, click the Create Revolve button or press R or right-click in the graphics window and select Create Feature > Revolve from the menu. The Revolve dialog box appears, as shown in the following image on the right.

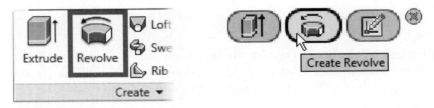

FIGURE 3.25

After starting the Revolve command, the in-canvas display options will appear in the graphics window and the Revolve dialog box will appear. The following image on the left shows the mini-toolbar revolve options that are available when revolving the first sketch and the image on the right shows the operation button that is available when creating additional revolve features. You can either enter data in the dialog box or via the in-canvas display. For the in-canvas buttons, click the down arrow next to each button to see more options and click and drag the arrow to change the distance or taper. The in-canvas buttons will change depending upon what options are selected. If selection tags have a red arrow, that command option needs to be satisfied before the command can be completed.

FIGURE 3.26

The Revolve dialog box as shown in the following image has five sections: Shape, Operation, Extents, Output, and Match Shape. When you make changes in the dialog box, the preview for the revolved feature changes in the graphics area to represent the values and options selected. When you have entered the values and the options that you need, click the OK button.

FIGURE 3.27

Shape

This section has two options: Profile and Axis.

	Profile	Click this button to choose the profile to revolve. If the Profile button is shown depressed, this is telling you that a profile or sketch needs to be selected. If there are multiple closed profiles, you will need to select the profile that you want to revolve. If there is only one possible profile, Autodesk Inventor will select it for you, and you can skip this step. If the wrong profile or sketch area is selected, click the Profile button, and choose the new profile or sketch area. To remove a selected profile, hold the CTRL key and click the profile to remove.
	Axis	Click a straight edge, centerline, work axis, or origin axis about which the profile(s) should be revolved. See the section below on how to create a centerline and create diametric dimensions.
	Solid	If there are multiple solid bodies, click this button to choose the solid body(ies) to participate in the operation.

Operation

This is the unlabeled middle column of buttons. If this is the first sketch that you create a solid from, it is referred to as a base feature, and only the top button is available. The operation defaults to Join (the top button). Once the base feature has been established, you can then revolve a sketch, adding or removing material from the part using the Join or Cut options, or you can keep what is common between the existing part and the completed revolve operation using the Intersect option.

	Join	Adds material to the part.
	Cut	Removes material from the part.
	Intersect	Removes material, keeping what is common to the existing part feature(s) and the new feature.
	New Solid	Creates a new solid body. The first solid feature created uses this option by default. Select to create a new body in a part file with an existing solid body.

Extents

The Extents determines if the sketch will be revolved 360°, a specified angle, stop at a specific plane, or start and stop at specific planes.

FIGURE 3.28

Full
Full is the default option; it will revolve the sketch 360° about a specified edge or axis.

Angle
Click this option from the drop-down list, and the Revolve dialog box displays additional options, as shown in the following image. Enter an angle for the sketch to be revolved. The three buttons below the degree area will determine the direction of the revolution. Choose the first two to flip the revolution direction, or click the right-side button to have the revolution go an equal distance in the negative and positive directions. If the angle was set to 90°, for example, the revolution will go 45° in both the negative and positive directions.

To Next
This option determines that the sketch will be revolved until it reaches a plane or face. The sketch must be fully enclosed in the area to which it is projecting; if it is not fully

enclosed, use the To termination with the Extend to Surface option. Click the Direction button to determine the revolve direction.

To

This option determines that the sketch will be revolved until it reaches a selected face, plane, or point. To select a plane, face, or point to end the revolve, click the Select Surface button, as shown in the following image, and then click a face, plane, or point at which the revolve should terminate.

Between

This option determines that the revolve will start at a selected plane or face and stop at another plane or face. Click the Select surface to start feature creation button, and then click the face or plane where the revolve will start. Click the Select surface to end the feature creation button, and then click the face or plane where the extrusion or revolve will terminate.

FIGURE 3.29

Direction

When the Extents is set to Angle there are four buttons from which to choose for determining the direction. Choose from the first two to flip the angle direction, click the third button (symmetric) to have the revolution go equal distances in the negative and positive directions, and click the asymmetric option to define different angles for the negative and positive direction. With the symmetric option, if the angle is 90 degrees, for example, the angle will go 45 degrees in both the negative and positive Z directions.

FIGURE 3.30

Output

Two options are available to select the type of output that the Revolve command will generate.

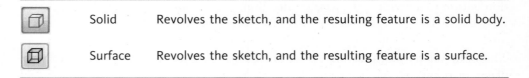

| | Solid | Revolves the sketch, and the resulting feature is a solid body. |
| | Surface | Revolves the sketch, and the resulting feature is a surface. |

Match Shape

You can use the Match shape option when working with an open profile. The edges of the open profile are extended until they intersect the geometry of the model. This provides a "flood-fill"–type effect for the revolve feature.

More

The More tab, as shown in the following image, contains additional options to refine the feature being created.

Alternate Solution

Alternate Solution terminates the feature on the most distant solution for the selected surface. An example is shown in the following image.

FIGURE 3.31

Minimum Solution

Minimum Solution terminates the feature on the first possible solution for the selected surface. An example is shown in the following image.

FIGURE 3.32

Infer iMates

Check this box to automatically create an iMate on a full circular edge. Autodesk Inventor attempts to place the iMate on the closed loop most likely to be used.

EXERCISE 3-2: REVOLVING A SKETCH

In this exercise, you will create a sketch and then create a revolved feature to complete a part. This exercise demonstrates how to revolve sketched geometry about an axis to create a revolved feature.

1. Click the New command.
2. Select the English tab, and then double-click Standard (in).ipt.
3. Create the sketch geometry as shown. Place the lower endpoint of the centerline at the projected origin point. Then use the Centerline command on the Format panel to change the left vertical line to a centerline as shown in the following image on the left.
4. Click the Dimension command.
5. Add the linear diameter dimensions shown in the following image on the right by selecting the centerline and selecting an endpoint on the sketch. Then click a point to place the dimension.

FIGURE 3.33

6. Right-click in the graphics window, and click Finish Sketch from the menu.
7. From the Create panel bar, click the Revolve command or click on the sketch mini-toolbar and click the Create Revolve button. Since there is only one possible closed profile, the profile is selected for you. Since there is only one centerline the centerline is also automatically selected as the axis.
8. Change the extents to Angle.
9. Enter **45 deg** or drag the arrow in the mini-toolbar display area until **45.00 degrees** is displayed. The preview of the model updates to reflect the change as shown in the following image.

FIGURE 3.34

10. Change the direction of the revolve to go counter clockwise by selecting the left flip button in the dialog box or click and drag the arrow in the graphics screen to go **45.00 degrees** in the opposite direction. The preview image will reverse the direction counter clockwise.
11. Set the revolve direction to symmetric.
12. Enter **90 deg** for the angle. The preview should resemble the following image.

FIGURE 3.35

13. In the mini-toolbar display area, change the extent to asymmetric as shown in the following image.
14. In the graphics window, click and drag the arrows to different values as shown in the following image.

FIGURE 3.36

15. Change the extents type to Full.
16. Click the OK button or click the green check mark in the mini-toolbar to create the feature.
17. Your part should resemble the following image.

FIGURE 3.37

18. Close the file. Do not save changes. End of exercise.

EDITING A FEATURE

After you create a feature, the feature consumes all of the dimensions that were visible in the sketch. If you need to change a feature's options and values, such as operation, extents, distance, taper etc. you edit the feature. You can also edit the feature's sketch and change the values of the dimensions or add or remove geometry. Editing a feature and feature's sketch are covered in the next sections.

Editing Feature Options

To change a feature's options, you need to edit the feature. There are multiple methods that you can use to edit the feature. There is no preferred method; use the method that works best for your workflow.

- While not in a sketch, move the cursor over the feature in the graphics window and click. Click the Edit "Feature Name" button from the mini-toolbar. The

following image on the left shows the mini-toolbar button for editing an extrusion.
- Double-click the features name in the browser; the dialog box that was used to create the feature will appear.
- In the browser, right-click the feature's name and select Edit Feature from the menu as shown in the following image on the right.

FIGURE 3.38

When editing a feature, the dialog box will appear that was used to create the feature. While editing you can change the feature's values and its options except the join operation on a base feature (first feature in the browser) and the output (solid or surface). The following image on the left shows the menu that appears after right-clicking on Extrusion1.

Editing a Feature's Sketch

In the last section, you learned how to edit the dimensions and the settings in which the feature was created. In this section, you will learn how to add and delete constraints, dimensions, and geometry in the original 2D sketch. To edit the 2D sketch of a feature, do the following:

- While not in a sketch, move the cursor over the feature in the graphics window whose sketch you want to edit and click. Then click the Edit Sketch button from the mini-toolbar as shown in the following image on the left.
- In the browser, expand the feature, right-click the name of the feature or sketch, and click Edit Sketch from the menu as shown in the following image on the right.
- In the browser, expand the feature and double-click on the sketch icon.
- Another method to change the value of dimensions on a sketch is to right-click the feature's name in the browser, and select Show Dimensions from the menu.

FIGURE 3.39

While editing the sketch, you can both add and remove objects. You can add geometry lines, arcs, circles, and splines to the sketch. To delete an object, right-click it and select Delete from the menu, or click it and press the DELETE key. If you delete an object from the sketch that has dimensions associated with it, the dimensions are no longer valid for the sketch, and they will also be deleted. You can also delete the entire sketch and replace it with an entirely new sketch. When replacing entire sketches, you should first delete other features that would be consumed by the new objects and re-create them.

You can also add or delete constraints and dimensions. When the dimensions are visible on the screen, double-click the dimension text that you want to edit. The Edit Dimension dialog box appears. Enter a new value, and then click the checkmark in the dialog box or press the ENTER key. Continue to edit the dimensions and, when finished click Finish Sketch from the ribbon or click the Local Update button on the Quick Access toolbar as shown in the following image. The sketch will be consumed by the feature and the new values will be used to regenerate the part.

FIGURE 3.40

If you receive an error after updating the part, make sure that the sketch forms a closed profile. If the appended or edited sketch forms multiple closed profiles, you will need to reselect the profile area.	**NOTE**

Renaming Features and Sketches

By default, each feature is given a name. These feature names may not help you when trying to locate a specific feature of a complex part, as they will not be descriptive to your design intent. The first extrusion, for example, is given the name Extrusion1 by default, whereas the design intent may be that the extrusion is the thickness of a plate. To rename a feature, slowly double-click the feature name and enter a new name. Spaces are allowed.

Deleting a Feature

You may choose to delete a feature after it has been placed. To delete a feature, right-click the feature name in the browser, and select Delete from the menu, as shown in the following image. The Delete Features dialog box will then appear, and you should choose what you want to delete from the list. You can delete multiple features by holding down the CTRL or SHIFT key, clicking their names in the browser, right-clicking one of the names, and then selecting Delete from the menu. The Delete Features dialog box appears; it allows you to delete consumed or dependent sketches and features.

FIGURE 3.41

Failed Features

If the feature in the browser turns red and have a yellow triangle with an explanation point before the features icon as shown in the following image. After updating the part, this is an alert that the new values or settings were not regenerated successfully. To see what the feature looked like in its last successful state, move the cursor over the features name in the browser and it will highlight in the graphics window like what is shown in the following image on the right of the last good fillet feature. You can then edit the values, enter new values, or select different settings to define a valid solution. Once you define a valid solution, the feature should regenerate without error.

FIGURE 3.42

EXERCISE 3-3: EDITING FEATURES AND SKETCHES

In this exercise, you will edit a consumed sketch in an extrusion and update the part. In the next section, you learn how to create sketched features that were created in this model.

1. Open *ESS_E03_03.ipt* from the Chapter 03 folder.
2. In the browser, right-click Extrusion1, and Click Edit Sketch from the menu. The feature's sketch is displayed.
3. Double-click the .375 radius dimension. In the field, enter **.25**, and then press Enter or click the checkmark in the dialog box.
4. Double-click the 2.000 dimension, and change the value to **3** and then press Enter or click the checkmark in the dialog box. The sketch should resemble the following image on the left.
5. Click the Finish Sketch command on the Sketch tab > Exit panel. The feature is updated with the new values as shown in the following image on the right.

FIGURE 3.43

6. You now edit the extents method that Extrusion2 was created with. In the graphics window, click on the inside circular face and then click on the Edit Extrude mini-toolbar button as shown in the following image on the left. The Extrude dialog box and mini-toolbar buttons are displayed.
7. Change the Operation to Join as shown in the following image on the right or use the mini-toolbar buttons.
8. Change the Extents option to To and select the bottom face as shown in the following image. The extrusion will stop at this face no matter what dimensional changes occur to the part.

FIGURE 3.44

9. Click OK or the green check mark in the mini-toolbar display. The feature is updated, as shown in the following image on the left.
10. In the browser, right-click Extrusion3. Click Delete and then click OK in the dialog box to delete consumed sketches and features. When done, your part should resemble the following image on the right.

FIGURE 3.45

11. Practice editing the sketch dimensions and features.
12. Close the file. Do not save changes. End of exercise.

SKETCHED FEATURES

A sketched feature is created from a sketch that you draw on a planar face or work plane and then add or remove material from the part. The basic steps to create a sketched feature are as follows:

1. Create a 2D sketch or make an existing sketch active.
2. Draw the geometry that defines the sketch.
3. Add constraints and dimensions.
4. Perform a boolean operation that will either add material to the part, remove material from the part, or keep whatever is common between the part and the feature that is being created. The following image shows a sketch on the left and the affect the three operations would have on the part.

FIGURE 3.46

There are no limits to the number of sketched features that can be added to a part. Each sketched feature is created on its own plane, and multiple features can reference the same plane. You can sketch curves, apply constraints, and apply dimensions exactly as you did with the first sketch. In addition to constraining and dimensioning the new sketch, you can also constrain the new sketch to the existing part. You can place dimensions to geometry that does not lie on the current plane; the dimensions, however, will be placed on the current plane. After the sketch has been constrained and dimensioned, it can be extruded, revolved, swept, or lofted. Exit the sketch environment by right-clicking in the graphics area and selecting Finish Sketch from the menu or by clicking the Finish Sketch command on the Exit panel. Only one sketch can be active at a time.

In the following section, you will learn how to assign a plane to the active sketch and then how to work with sketched features.

DEFINE THE ACTIVE SKETCH PLANE

As stated previously, each sketch must exist on its own plane. The active sketch has a plane on which the sketch is drawn. To assign a plane to the active sketch, the plane on which the sketch will be created must be a planar face, a work plane, or an origin plane. The planar face does not need to have a straight edge. A cylinder has two planar faces, one on the top and the other on the bottom of the part. Neither has a straight edge, but a sketch can be placed on either face.

To make a sketch active, use one of the following methods:

- Click the 2D Sketch command from the Model tab > Sketch panel as shown in the following image on the left, and then click the plane where you want to place the sketch.
- While not in a sketch, click on the planar face or workplane that you want to create the active sketch on. Then click the Create Sketch button from the mini-toolbar buttons as shown in the following image on the right.
- Press the key S, and then click the plane where you want to place the sketch.
- Click a plane that will contain the active sketch, and press the hot key S.
- Click a plane that will contain the active sketch, right-click in the graphics area, and select New Sketch from the menu.
- Issue the Sketch command, expand the Origin folder in the browser, and click one of the origin planes.
- Expand the Origin folder in the browser, right-click on one of the default planes, and select New Sketch on the menu.
- To make a previously created sketch active, issue the Create 2D Sketch command and click the sketch name in the browser.

Once you have created a sketch, it appears in the browser with the name Sketch#, and sketch commands appear in the 2D Sketch panel bar. The number will sequence for each new sketch that is created. To rename a sketch, go to the browser, slowly double-click the existing name, and enter the new name. When you create (PE) a new sketch or make a sketch active, by default the view will automatically change so you are looking straight at the sketch. This is controlled by the Application Option > Sketch tab > Look at sketch plane on sketch creation plan view, that is, looking straight at, or normal to, the sketch plane. This can be done by using the View Face command and click any geometry on the sketch or click the sketch in the browser.

FIGURE 3.47

Select Other-Face Cycling

Autodesk Inventor has dynamic face highlighting that helps you to select the correct face, edge, part etc. to activate and to select objects. As you move the cursor over a given face, the edges of the face are highlighted. If you continue to move the cursor, different faces are highlighted as the cursor passes over them.

To cycle to a face that is behind another one, start a command and then move the cursor over the face that is in front of one that you want to select and hold the cursor still for two seconds. The Select Other tool appears, as shown in the following image on the left. Select the drop down arrow and move the cursor over the available objects in the list until the correct face is highlighted and then click. The number of objects that appear in the list will depend upon the geometry and the location of the cursor, when the Select Other tool appears. You can also access the Select Other command by right-clicking the desired location in the graphics area and clicking Select Other from the marking menu.

You can specify the amount of time before the Select Other command will appear automatically. Click Application Options on the Tools tab > Options panel and click the General tab. In the Selection area as shown in the following image on the right, the "Select Other" delay (sec) feature can be specified in tenths of a second. If you do not want the Select Other command to open automatically, specify OFF in the field. The default value is 1.0 second.

FIGURE 3.48

SLICE GRAPHICS

While creating parts, you may need to sketch on a plane that is difficult to see because features are obscuring the view. The Slice Graphics option will temporarily slice away the portion of the model that obscures the active sketch plane on which you want to sketch. The following image on the left shows a revolved part with an origin plane visible and the Slice Graphics menu. The image on the right shows the graphic sliced and the origin plane's visibility turned off. To temporarily slice the graphics screen, follow these steps:

1. Make a planar face or work plane that the graphics of the active sketch will be sliced through.
2. Rotate the model so the correct side will be sliced, that is, the side of the model that faces up will be sliced away.
3. While editing the sketch, right-click and select Slice Graphics from the menu as shown in the following image, press the F7 key or click Slice Graphics on the View tab > Appearance panel. The part will be sliced on the active sketch plane.
4. Use sketch commands from the Sketch tab to create geometry on the active sketch.
5. To exit the sliced graphics environment, right-click and select Slice Graphics, press the F7 key or click the Finish Sketch button on the Exit panel.

FIGURE 3.49

> **NOTE**
> When working in an assembly, you have additional Assembly Section commands available from the View tab > Appearance panel.

EXERCISE 3-4: SKETCH FEATURES

In this exercise, you will create a sketch on the angled face of a part and then create an oblong on the new sketch plane.

1. Open *ESS_E03_04.ipt* from the Chapter 03 folder.
2. Click the Create 2D Sketch command on the Sketch panel, and then click the top-inside angled face as shown in the following image on the left. You could have clicked on the face and click the Create Sketch mini-toolbar button.
3. The view should change so you are looking straight at the sketch, if not click the View Face command from the Navigation bar, and then select Sketch6 in the browser.

4. Next you create a sketch for an obround in the part and center the slot vertically. Sketch a slot using dynamic input to create a **1.5 inch** high obround as shown in the following image on the right. Both arcs should be tangent to the adjacent lines.

FIGURE 3.50

5. Click the Horizontal constraint command on the Constrain panel as shown in the following image on the left.
6. Select the midpoint of the right vertical edge of the oblong and the midpoint of the right vertical edge of the part as shown in the following image on the right. The obround is centered vertically using the horizontal constraint.

FIGURE 3.51

7. Click the Dimension command on the Constrain panel as shown in the following image on the left. To fully constrain the sketch, place the two dimensions as shown in the following image on the right.

FIGURE 3.52

8. Click the Finish Sketch command on the Exit panel.
9. If the view did not automatically change to the home view, press the F6 key to change to the Home View.
10. In the graphics window, click on an entity on the obround and click Create Extrude from the mini-toolbar buttons.
11. For the profile, click inside the oblong.
12. In the mini-tool, click the Cut operation, and change the Extents to All. Ensure that the direction is pointing into the part as shown in the following image on the left.
13. Click OK. The completed part is shown in the following image on the right.

FIGURE 3.53

14. Practice creating sketched features trying different options.
15. Close the file. Do not save changes. End of exercise.

PROJECTING PART EDGES

Building parts based partially on existing geometry is done often, and you will frequently need to reference faces, edges, or loops from features that have been created.

While in a part file, you can project an edge, face, point, or loop onto a sketch. Projected geometry can maintain an associative link to the original geometry that is projected. If you project the face of a feature onto another sketch, for example, and the parent sketch is modified, the projected geometry will update to reflect the changes.

Direct Model Edge Referencing

While you sketch, you can use direct model edge referencing to:

- Automatically project edges of the part to the sketch plane as you sketch a curve.
- Create dimensions and constraints to edges of the part that do not lie on the sketch plane.
- Control the automatic projection of part edges to the sketch plane.

Creating Reference Geometry

There are two ways to automatically project part edges to the sketch plane:

- Click an edge of the part while creating a dimension or constraint.

- On the Sketch tab in the Application Options dialog box, you can control two options:
  - Autoproject edges during curve creation, which controls the ability to touch and project edges while sketching a curve.
  - Autoproject edges for sketch creation and edit, which controls the automatic projection of the edges of the selected face when you start a sketch on a planar face of the part.

NOTE Neither of these options disables the ability to reference part edges when creating dimensions and constraints.

PROJECT EDGES

In this section, you learn about using the Project Geometry command that can project selected edges, vertices, work features, curves, the silhouette edges of another part in an assembly, or other features in the same part to the active sketch. There are three project commands available on the Sketch Panel Bar: Project Geometry, Project Cut Edges, and Project Flat Pattern, as shown in the following image.

Project Geometry
Use to project geometry from a sketch or feature onto the active sketch.

Project Cut Edges
Use to project part edges that touch the active sketch. The geometry is only projected if the uncut part would intersect the sketch plane. For example, if a sphere has a sketch plane in the center of the part and the Project Cut Edges command is initialized, a circle will be projected onto the active sketch.

Project Flat Pattern
Use to project a selected face or faces of a sheet metal part flat pattern onto the active sheet metal part sketch plane.

Project to 3D Sketch
Use to project geometry from the active 2D sketch onto selected faces to create a 3D sketch.

FIGURE 3.54

To project geometry, follow these steps:

1. Make a plane in the active sketch to which the geometry will be projected.
2. Click the Project Geometry command on the 2D Sketch Panel Bar.
3. Select the geometry to be projected onto the active sketch. Click a point in the middle of a face, and all edges of the face will be projected onto the active sketch plane. If you want to project all edges that are tangent to an edge (a loop), use the Select Other command to cycle through until they all appear highlighted, as shown in the following image on the left. The following image on the right shows the projected geometry.
4. To exit the operation, press the ESC key or click another command.

FIGURE 3.55

If you clicked a loop for the projection, the sketch is updated to reflect the modification when any part of the profile changes. If a face is projected, the internal islands that are defined on the face are also projected and will update accordingly. For example, if the loop of edges of the following image on the left is projected, the four lines that define the cutout will be projected and updated if the original feature is modified. To disassociate projected geometry from the original geometry, show the sketch constraints and delete the reference constraint that is created for each projected edge/curve.

FIGURE 3.56

EXERCISE 3-5: PROJECT GEOMETRY

In this exercise, you will project geometry to create a feature, edit the original sketch and update the linked feature and then delete the sketch constraints that were created from the projected geometry.

1. Open *ESS_E03_05.ipt* from the Chapter 03 folder.
2. Create a sketch on the bottom face of the cylinder (opposite side of Extrusion2) as shown in the following image on the left.
3. Rotate your viewpoint so you can see the front of the part.
4. Click the Project Geometry command on the Draw panel and select the front face the Extrusion2 as shown in the following image on the right.

FIGURE 3.57

5. Again rotate your viewpoint so you can see the bottom of the part. Notice that the square has been projected onto the active sketch and is fully constrained.
6. Finish the sketch.
7. Extrude the sketch **.5** inches, cutting material into the part as shown in the following image on the left.
8. Edit the sketch of Extrusion2 (the original extruded square).
9. Change the 1 in dimension to **.5** in as shown in the following image on the right.

FIGURE 3.58

10. Finish the sketch and rotate the viewpoint so you can see the bottom of the part. Notice that Extrusion3's sketch and feature have been updated to reflect the change to the geometry that was projected.
11. Edit the sketch of Extrusion3.

12. Next you will delete the reference constraints that were created when the geometry was projected. Press the F8 key to display the constraints as shown in the following image on the left.
13. Delete the four reference constraints that surround the square (these constraints were automatically created when the geometry was projected onto this sketch). The sketch should now have 8 dimensions/constraints required to fully constrain the sketch.
14. Press the F8 key to refresh the constraints that are displayed, the only constrains that remain on the sketch are the four coincident constraints.
15. Click the Automatic Dimension command from the Constrain panel, select the four inside lines for the curves to Auto Dimension, uncheck the Dimension option (only constraints will be applied) as shown in the following image on the right and then click Apply in the dialog box.

FIGURE 3.59

16. Click the Done button in the dialog box.
17. To display the constraints that were added in the last step, press the F8 key.
18. Place a horizontal constraint to the bottom line.
19. Press the F9 key to hide the constraints.
20. To fully constrain the sketch, add the four dimensions as shown in the following image in the middle.
21. Finish the sketch and rotate the part so you see that Extrusion3 is a different size than Extrusion2 as shown in the following image on the right.

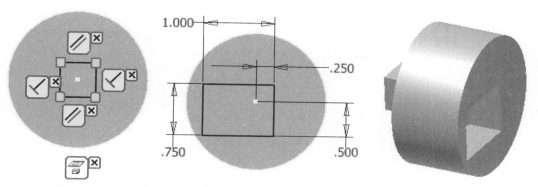

FIGURE 3.60

22. Close the file. Do not save changes. End of exercise.

PART MATERIAL, PROPERTIES, AND COLOR

To change a part's physical properties and appearance to a specific material, click the Inventor Application Menu and click iProperties from the menu. You can also right-click on the part's name in the browser and click iProperties from the menu or click Physical Properties from the Color list on the Quick Access toolbar. Then in the iProperties dialog box click the Physical tab and select a material from the material drop-down list as shown in the following image on the left. Click the Update button in the dialog box to update the properties of the file as shown in the following image on the right.

FIGURE 3.61

Click the OK button in the dialog box to complete the operation, and the color of the part change to reflect the selected material. If the part color does not update when the material is edited, click the As Material option in the Color area of the Quick Access toolbar, as shown in the following image on the right. If you selected a color from the Color drop-down list on the Quick Access toolbar, it will change only the color of the part, not the physical material or properties of the part.

FIGURE 3.62

APPLYING YOUR SKILLS

Skill Exercise 3-1

In this exercise, you will create a bracket from a number of extruded features. Assume that the part is symmetrical about the center of the horizontal slot.

1. Start a new part based on the English *Standard (in).ipt* template.
2. Create the sketch geometry for the base feature.
3. Add geometric constraints and dimensions. Make sure that the sketch is fully constrained.
4. Extrude the base feature.
5. Create the three remaining features to complete the part.

FIGURE 3.63

The completed part should resemble the following image. When done close the file. Do not save changes. End of exercise.

FIGURE 3.64

Skill Exercise 3-2
In this exercise, you will create a connecting rod and add draft to extrusions during feature creation.

1. Start a new part based on the English *Standard (in).ipt* template.
2. Create the sketch geometry for the outside of the connecting rod.
3. Add geometric constraints and dimensions to fully constrain the sketch.
4. Extrude the base feature using the midplane option, adding a −10° taper.
5. Create a separate feature for each pocket by projecting geometry (don't mirror the feature, the mirror command will be covered in Chapter 7). The sides of the pocket are parallel to the sides of the connecting rod.

FIGURE 3.65

The completed part should resemble the following image. When done close the file. Do not save changes. End of exercise.

FIGURE 3.66

Skill Exercise 3-3
In this exercise, you will create a pulley using a revolved feature. Assume that the part is symmetric about the middle of the part vertically.

1. Start a new part based on the English *Standard (in).ipt* template.
2. Create the sketch geometry for the cross-section of the pulley.

To create a centerline, draw a line, select it, and then click the Centerline command on the Sketch tab > Format panel.	TIP

3. Apply appropriate sketch constraints.
4. Add dimensions.
5. Revolve the sketch.

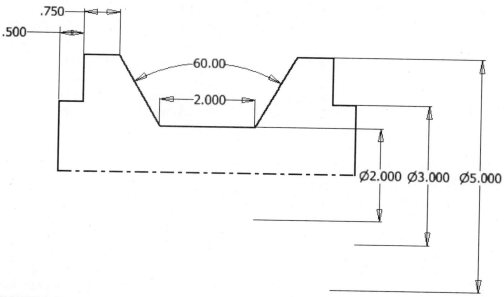

FIGURE 3.67

The completed part should resemble the following image. When done, close the file. Do not save changes. End of exercise.

FIGURE 3.68

CHECKING YOUR SKILLS

Use these questions to test your knowledge of the material covered in this chapter.

1. What is a base feature?
2. True ___ False ___ When creating a feature with the Extrude or Revolve command, the mini-toolbar buttons only allows you to define the distance or angle.
3. When creating a revolve feature, which objects can be used as an axis of revolution?
4. Explain how to create a linear diameter (diametric) dimension on a sketch.
5. Name two ways to edit an existing feature.
6. True ___ False ___ Once a sketch becomes a base feature, you cannot delete or add constraints, dimensions, or objects to the sketch.
7. Name three operation types used to create sketched features.
8. True ___ False ___ A direct manipulation technique can only be started by clicking on a face of a part.
9. True ___ False ___ Once a sketched feature exists, its extents type cannot be changed.
10. True ___ False ___ By default geometry that is projected from one feature to a sketch will update automatically based on changes to the original projected geometry.
11. Explain what the asymmetric option does for the extrude and revolve commands.
12. Where do you set the physical material property of a part?
13. True ___ False ___ After setting a part's material properties, the color setting in the Quick Access toolbar must be set to Color Matching for the color to match the material of the part.
14. True ___ False ___ The Project Geometry command is used to copy objects onto any selected face.
15. True ___ False ___ By default, when a feature is deleted, the feature's sketch will be maintained.

CHAPTER 4

Creating Placed Features

INTRODUCTION

In Chapter 3, you learned how to create and edit base and sketched features. In this chapter, you will learn how to create *placed* features. These are features that are predefined except for specific values and only need to be located. You can edit placed features in the browser like sketched features. When you edit a placed feature, either the dialog box that you used to create it will open or feature values will appear on the part.

When creating a part, it is usually better to use placed features instead of sketched features wherever possible. To make a through hole as a sketched feature, for example, you can draw a circle profile, dimension it, and then extrude it with the cut operation, using the All extension. You can also create a hole as a placed feature—you can select the type of hole, size it, and then place it using a dialog box. When drawing views are generated, the type and size of the hole are easy to annotate, and they automatically update if the hole type or values change.

OBJECTIVES

After completing this chapter, you will be able to perform the following:

- Create fillets
- Create chamfers
- Create holes
- Shell a part
- Create work axes
- Create work points
- Create work planes
- Create a UCS
- Pattern features

FILLETS

Fillet features consist of fillets and rounds. Fillets add material to interior edges to create a smooth transition from one face to another. Rounds remove material from exterior edges. The following image shows a part without fillets on the left and the part with fillets on the right.

FIGURE 4.1

To create fillets, you select the edge that needs to be filleted; the fillet is created between the two faces that share the edge, or you can select two or three faces that a fillet will go between. When placing a fillet between two faces, the faces do not need to share a common edge. When creating a part, it is good practice to create fillets and chamfers as some of the last features in the part. Fillets add complexity to the part, which in turn adds to the size of the file. They also remove edges that you may need to place other features.

To create a fillet feature, click the Fillet command from the Model tab > Modify panel, as shown in the following image on the left, while not in a sketch click on an edge and the mini-toolbar will appear click the Create Fillet button as shown in the following image on the right or press the shortcut key F.

FIGURE 4.2

After you click the command, the Fillet dialog box appears as shown in the following image on the left and the mini-toolbar will appear as shown in the following image on the right. Along the left side of the Fillet dialog box, there are three types of fillets: Edge, Face, and Full Round. When the Edge option is selected, you see three tabs: Constant, Variable, and Setbacks. Each tab creates a fillet along an edge(s) with different options. The options for each of the tabs and fillet types are described in the following sections. A preview of a fillet will appear on the part when the preview option on the bottom of the dialog box is checked and a valid fillet can be created from the data you input into the Fillet dialog box or the mini-toolbar.

FIGURE 4.3

Before we look at the tabs, let's discuss the methodology that you will use to create fillets. You can click an edge or edges, or faces to fillet, or select the type of fillet to create. If you click an edge or face before you issue the Fillet feature command, it is placed in the first selection set. Selection sets contain the edges or faces that will be filleted when the OK button is clicked.

Each fillet feature can contain multiple selection sets, each having its own unique fillet value. There is no limit to the number of selection sets that can exist in a single instance of the feature. An edge, however, can only exist in one selection set. All of the selection sets included in an individual fillet command appear as a single fillet feature in the browser. Click the type of fillet that you want to create. To add edges to the first selection set, select the edges that you want to have the same radius. To create another selection set, select Click to add, and then click the edges that will be part of the next selection set. To remove an edge or face that you have selected, click the selection set that includes the edge or face, and the edges or faces will be highlighted. Hold down the CTRL key, and click the edge(s) or face(s) to be removed from the selection set. Enter the desired values for the fillet. As changes are made in the dialog box, a representation of the fillet is previewed in the graphics area. When the fillet type and value are correct, click the OK button to create the fillet.

To start editing a fillet's type and radius, use one of the following methods to start the editing process:

- In the graphics window, click on the fillet to edit and click the Edit Fillet button that appears in the graphics window.

- Start the Edit Feature command by right-clicking on the fillet's name in the browser and selecting Edit Feature from the menu.
- Double-click on the feature's name in the browser.
- Change the select priority to Select Feature and double-click on the fillet that is on the part.

The Fillet dialog box appears with all of the settings that you used to create the fillet. Change the fillet settings as needed. Then if needed click the Local Update command on the Quick Access toolbar to update the part.

Edge Fillet

With the Edge fillet type selected, the Constant tab will be activated. With the options on the Constant tab, as shown in the previous image, you can create fillets that have the same radius from beginning to end. There is no limit to the number of part edges that you can fillet with a constant fillet. You can select the edges as a single set or as multiple sets, and each set can have its own radius value. The order in which you select the edges of a selection set is not important. You need to select the edges that are to be filleted individually, and the use of the window or crossing selection method is not allowed. If you change the value of a selection set, all of the fillets in that group will change. To remove an edge from a group, choose the group in the select area, and then hold down the CTRL key and click the edge to be removed.

Constant Tab

The following section describes the options that are available on the Constant tab.

Select Edge, Radius, and Continuity

Edge

By default, after issuing the Fillet command, you can click edges, and they appear in the first selection set. You can continue to select multiple edges. To remove an edge from the set, hold down the CTRL key and select the edge.

Radius

Enter a size for the fillet. The size of the fillet will be previewed on the selected edges.

Continuity

To adjust the continuity of the fillet, select either a tangent or smooth (G2) condition (continuous curvature), as shown in the following image in both the dialog box and in the mini-toolbar.

FIGURE 4.4

To create another selection set, select Click to add, and then click the edges that will be part of the next selection set. After clicking an edge, a preview image of the fillet appears on the edge that reflects the current values that are set in the dialog box or mini-toolbar, as shown in the following image.

FIGURE 4.5

Select Mode

Edge
Click the Edge mode to select individual edges to fillet. By default, any edge that is tangent to the clicked edges is also selected. If you do not want to have tangent edges automatically selected, uncheck the Automatic Edge Chain option in the More (>>) section.

Loop
Click the Loop mode to have all of the edges that form a closed loop with the selected edge filleted.

Feature
Click the Feature mode to select all of the edges of a selected feature.

Solids
If multiple solid bodies exist, select the solid body to apply All Fillets or All Rounds.

All Fillets
Click the All Fillets option to select all concave edges of a part that you have not filleted already—see the following image on the left. The All Fillets option adds material to the part, and it requires a separate edge selection set to remove material from the remaining edges using the All Rounds option.

All Rounds
Click the All Rounds option to select all convex edges of a part that you have not filleted already, as seen in the following image on the right. The All Rounds option removes material from the part, and it requires a separate edge selection set from All Fillets.

FIGURE 4.6

More (>>) Options. Click the More (>>) button to access other options, as shown in the following image.

Roll along Sharp Edges
Click this option to adjust the specified radius when necessary to preserve the edges of adjacent faces.

Rolling Ball Where Possible
Click this option to create a fillet around a corner that looks like a ball has been rolled along the edges that define a corner, as shown in the following (middle) image. When the Rolling ball where possible solution is possible, but it is not selected, a blended solution is used, as shown in the following image on the right.

FIGURE 4.7

Automatic Edge Chain
Click this option to select tangent edges automatically when you click an edge.

Preserve All Features
Click to check all features that intersect with the fillet and to calculate their intersections during the fillet operation. If the option's checkbox is clear, only the edges that are part of the fillet operation are calculated during the operation.

For example, if Preserve All Features is checked, a fillet is placed on the outside edge of a shelled box. Since the fillet is larger than the shell thickness, a gap will exist in the fillet, as shown in the image on the left. If the Preserve All Features is unchecked, the inside edges for the shell will appear in the fillet, as shown in the following image on the right. The default option is NOT to preserve the features.

FIGURE 4.8

Face Fillet

With the Face fillet type selected, the dialog box will change, as shown in the following image on the left and the mini-toolbar on the right. Create a face fillet by selecting two or more faces; the faces do not need to be adjacent. If a feature exists that will be consumed by the fillet, the volume of the feature will be filled in by the fillet.

FIGURE 4.9

To create a face fillet, follow these steps:

1. With the Face Set 1 button active, select one or more tangent contiguous faces on the part to which the fillet will be tangent.
2. With the Face Set 2 button active, select one or more tangent contiguous faces on the part to which the fillet will be tangent.

Check the Include Tangent Faces option to automatically chain all faces that are tangent to faces in the selection set.

Check the Optimize for Single Selection to automatically make the next selection set button active after selecting a face.

The following image on the left shows a part that has a gap between the bottom and top extrusion, the middle image shows the preview of the face fillet, and the image on the right shows the completed face fillet. Notice the rectangular extrusion on the top face is consumed by the fillet.

FIGURE 4.10

FullRound Fillet

With the FullRound fillet type selected, the dialog box will change, as shown in the following image on the left and the mini-toolbar on the right. The full round fillet option creates a fillet that is tangent to three faces; the faces do not need to be adjacent.

FIGURE 4.11

To create a FullRound fillet, follow these steps:

1. With the Side Face Set 1 button active, select a face on the part that the fillet will start at and to which it will be tangent.
2. With the Center Face Set button active, select a face on the part that the middle of the fillet will be tangent.
3. With the Side Face Set 2 button active, select a face on the part that the fillet will end at and to which it will be tangent.

Check the Include Tangent Faces option to automatically chain all faces that are tangent to faces in the selection set.

Check the Optimize for Single Selection to automatically make the next selection set button active after selecting a face.

The following image shows three faces selected on the left and the resulting fullround fillet on the right.

FIGURE 4.12

> If you get an error when creating or editing a fillet, try to create it with a smaller radius. If you still get an error, try to create the fillet in a different sequence or create multiple fillets in the same operation.

TIP

CHAMFERS

Chamfers are similar to fillets except that their edges are beveled rather than rounded. When you create a chamfer on an interior edge, material is added to your model. When you create a chamfer on an exterior edge, material is cut away from your model, as shown in the following image.

FIGURE 4.13

To create a chamfer feature, follow the same steps that you used to create fillet features. Click the common edge, and the chamfer is created between the two faces sharing the edge. To create a chamfer feature, click on the Chamfer command on the Model tab > Modify panel, as shown in the following image on the left, or press the keys CTRL SHIFT K. Another method, while not in a sketch, click on an edge and the mini-toolbar will appear, then click the Create Chamfer button in the mini-toolbar, as shown in the following image on the right.

FIGURE 4.14

After you start the command, the Chamfer dialog box appears, as shown in the following image on the left and the mini-toolbar appears as shown in the following image on the right. As with fillet features, you can select multiple edges to be included in a single chamfer feature. From the dialog box or mini-toolbar, click a method, click the edge or edges to chamfer, enter a distance and/or angle, and then click OK.

FIGURE 4.15

To edit the type of chamfer feature or distances, use one of the following methods to start the editing process:

- In the graphics window, click on the chamfer to edit and click the Edit Chamfer button that appears in the mini-toolbar.
- Double-click the feature's name or icon in the browser to edit a distance.
- Right-click the chamfer's name in the browser, and select Edit Feature from the menu. The Chamfer dialog box appears with all the settings that you used to create the feature. Adjust the settings as desired.
- Change the select priority to Feature Priority, and then double-click the chamfer on the part. The Chamfer dialog box appears with all the settings that you used to create the feature. Adjust the settings as desired.
- The Chamfer dialog box appears with all of the settings that you used to create the fillet. Change the chamfer settings as needed. Then if needed click the Local Update command on the Quick Access toolbar to update the part.

Method

Distance

Click the Distance option to create a 45° chamfer on the selected edge. You determine the size of the chamfer by typing a distance in the dialog box. The value is the offset from the common edge of the two adjacent faces. You can select a single edge, multiple edges, or a chain of edges. A preview image of the chamfer appears on the part. If you select the wrong edge, hold down the CTRL key and select the edge to remove. The following image illustrates the use of the Distance option on the left.

Distance and Angle

Click the Distance and Angle option to create a chamfer offset from a selected edge on a specified face, at the defined angle. In the dialog box, enter an angle and distance for the chamfer, then click the face to which the angle is applied and specify an edge to be chamfered. You can select one edge or multiple edges. The edges must lie on the selected face. A preview image of the chamfer appears on the part. If you have selected the wrong face or edge, click on the Edge or Face button, and choose a new face or edge. The following image in the middle illustrates the use of the Distance and Angle option.

Two Distances

Click the Two Distances option to create a chamfer offset from two faces, each being the amount that you specify. Click an edge first, and then enter values for Distance 1 and Distance 2. A preview image of the chamfer appears. To reverse the direction of the distances, click the Flip button. When the correct information about the chamfer is in the dialog box, click the OK button. You can only use a single edge or chained edges with the Two Distances option. The following image illustrates the use of the Two Distances option on the right.

FIGURE 4.16

Edge and Face

Edges
Click an edge or edges to be chamfered.

Face
Click a face on which the chamfer will be based.

Flip
Click the button to reverse the direction of the distances for a Two Distances chamfer.

Distance and Angle

Distance
Enter a distance to be used for the offset.

Angle
Enter a value that will be used for the angle if creating the Distance and Angle chamfer type.

Edge Chain and Setback

Edge Chain
Click this option to include tangent edges in the selection set automatically, as shown in the following image.

Setback
When the Distance method is used and three chamfers meet at a vertex, click this option to have the intersection of the three chamfers form a flat edge (left button) or to have the intersection meet at a point as though the edges were milled (right button), as shown in the following image.

Preserve All Features

Click this option to check all features that intersect with the chamfer and to calculate their intersections during the chamfer operation, as shown in the following image. If the option's checkbox is clear, only the edges that are part of the fillet operation are calculated during the operation.

FIGURE 4.17

EXERCISE 4-1: CREATING FILLETS AND CHAMFERS

In this exercise, you will create constant radius fillets, full round fillets, face fillet, and chamfers.

1. Open *ESS_E04_01.ipt* in the Chapter 04 folder.
2. Click the Fillet command in the Model tab > Modify panel or click the inside edge of the cutout and click Create Fillet from the mini-toolbar.
3. In the Fillet dialog box, click on the first entry in the Radius column, and type **.25** or in the mini-toolbar enter **.25** as shown in the following image.

FIGURE 4.18

4. Click Apply to create the fillet.
5. Next, create a Full Round Fillet that will be tangent to the front three faces. In the Fillet dialog box, click the Full Round Fillet option in the left column.
6. Click the front-inside face, then the left-front face, and then the back-vertical face of the part, as shown in the following image.

FIGURE 4.19

7. Click OK to create the fillet.
8. In the browser, right-click on Extrusion4, and click Unsuppress Features.
9. Next, create a Face Fillet. Click the Fillet command in the Modify panel. In the Fillet dialog box, click the Face Fillet option in the left column.
10. Click the cylindrical face of Extrusion4 and then the full round fillet in the model you created in the last step, as shown in the following image.
11. In the Fillet dialog box, type **.375 in** in the Radius field, as shown in the following image.

FIGURE 4.20

12. Click OK to create the fillet and rotate the viewpoint to examine the fillet features.
13. Move the cursor into a blank area in the graphics window, right-click, and click Repeat Fillet from the menu.
14. Next you fillet two edges. Click the top and bottom edges of the model, as shown in the following image.
15. In the mini-toolbar input radius field, type **.125**, as shown in the following image.

FIGURE 4.21

16. Click the green check mark (OK button) in the mini-toolbar to create the fillets.
17. Click the back cylindrical edge on Extrusion4 and click Create Chamfer from the mini-toolbar as shown in the following image on the left.
18. In the mini-toolbar's distance field, type **.25**, as shown in the following image on the right.
19. To create the chamfer and keep the dialog box open, click green plus button in the mini-toolbar. The green plus button is the same as the Apply button in the dialog box.

FIGURE 4.22

20. In the Chamfer dialog box or in the mini-toolbar, click the Distance and Angle option.
21. Click the front-circular face on Extrusion4.
22. Click the front-circular edge of the same face.
23. In the Chamfer dialog box or in the mini-toolbar, enter **.25** in the Distance field and **60 deg** in the Angle field, as shown in the following image.

FIGURE 4.23

24. Click OK to create the chamfer. When done, your model should resemble the following image.

FIGURE 4.24

25. Practice editing and placing fillets and chamfers.
26. Close the file. Do not save changes. End of exercise.

HOLES

The Hole command lets you create drilled, counterbored, spotface, countersunk, clearance, tapped, and taper tapped holes, as shown in the following image. You can place holes using sketch geometry or existing planes, points, or edges of a part. You can also specify the type of drill point and thread parameters.

FIGURE 4.25

To create a hole feature, follow these steps:

1. Create a part that you want to place a hole on.
2. Click the Hole command from the Model tab > Modify panel, as shown in the following image on the left, or press the hot key H. You could also right-click and click Hole from the marking menu.
3. The Holes dialog box appears, as shown in the following image on the right. Four placement options are available: From Sketch, Linear, Concentric, and On Point. The Placement options are covered in the next section. After you have chosen the hole placement options, select the desired hole style options from the Holes dialog box. As you change the options, the preview image of the hole(s) is updated. When you are done making changes, click the OK button to create the hole(s).

FIGURE 4.26

Editing Hole Features

To edit the type of hole feature or distances, use one of the following methods:

- Double-click the feature's name or icon in the browser to display the Holes dialog box with the dimensions and option you used to create the hole feature.
- Change the select priority to Feature Priority, and then double-click the hole feature on the part in the graphics area. This will display the Holes dialog box with the dimensions and option you used to create the hole feature.
- Right-click the hole's name or icon in the browser, and select Edit Feature from the menu to display the Holes dialog box with the dimensions and option you used to create the hole feature.

Holes Dialog Box

In the Holes dialog box, you establish the placement method, type of hole, its termination, and additional options such as type of drill point, angle, and tapped properties.

Placement

Select the appropriate placement method. If you select From Sketch, a sketch containing hole centers or any point, such as an endpoint of a line, must exist on the part. Hole centers are described in the next section. The Linear, Concentric, and On Point options do not require an unconsumed or shared sketch to exist in the model and are based on previously created features. Depending on the placement option you select, the input parameters will change, as shown in the following images and as described next.

FIGURE 4.27

From Sketch

Select the From Sketch option to create holes that are based on a location defined within an unconsumed or shared sketch. You can base the center of the hole on a point/hole center or endpoints of sketched geometry like endpoints, centers of arcs and circles, and spline points. You can also use points from projected geometry that resides in the unconsumed or shared sketch.

Centers. Select the hole center point or sketch points where you want to create a hole.

Solids. If multiple solid bodies exist, select the solid body where you want to create a hole.

Linear. Select the Linear option to place the hole relative to two selected face edges.

Face. Select the face on the part where the hole will be created.

Solids. If multiple solid bodies exist, select the solid body where you want to create a hole.

Reference 1. Select a face edge as a positional reference for the center of the hole. When you select the edge, a dimension appears that can be edited to constrain the center of the hole dimensionally.

Reference 2. Select a face edge as a positional reference for the center of the hole. When you select the edge, a dimension appears that can be edited to constrain the center of the hole dimensionally.

Flip Side. Click this button to position the hole on the opposite side of the selected edge.

Concentric. Select the Concentric option to place the hole on a planar face and concentric to a circular or arc edge or a cylindrical face.

Plane. Select a planar face or plane where you want to create the hole.

Solids. If multiple solid bodies exist, select the solid body where you want to create the hole.

Concentric Reference. Select a circular or arc model edge or cylindrical face to constrain the center of the hole to be concentric with the selected entity.

On Point

Select the On Point option to place the center of the hole on a work point. The work point must exist on the model prior to selecting this option.

Point. Select a work point to position the center of the hole.

Solids. If multiple solid bodies exist, select the solid body where you want to create the hole.

Direction. Select a plane, face, work axis, or model edge to specify the direction of the hole. When selecting a plane or face, the hole direction will be normal to the face or plane.

Hole Option

Click the type of hole that you want to create: drilled, counterbore, spotface, or countersink, and enter the appropriate dimensions. The following image shows the counterbore option.

FIGURE 4.28

Termination

Select how the hole will terminate:

- *Distance:* Specify a distance for the depth of the hole.
- *Through All:* Choose to extend the hole through the entire part in one direction.
- *To:* Select a plane at which the hole will terminate.
- *Flip:* Reverse the direction in which the hole will travel.

Drill Point

Select either a flat or angle drill point. If you select an angle drill point, you can specify the angle of the drill point.

Infer iMates

Check this box to automatically create an iMate on a full circular edge. Autodesk Inventor attempts to place the iMate on the closed loop most likely to be used.

Dimensions

To change the diameter, depth, countersink, counterbore diameter, countersink angle, or counterbore depth of the hole, click the dimension in the dialog box, and enter a desired value.

Hole Type

Click the type of hole you want to create. There are four options: Simple Hole, Clearance Hole, Tapped Hole, and Taper Tapped Hole. The following image shows the tapped hole selected. After selecting the hole type, fill in the dialog box with the specific data for the hole you need to create.

FIGURE 4.29

Simple Hole. Click the Simple Hole option to create a (drilled) hole feature with no thread features or properties.

Clearance Hole. Click the Clearance Hole option to create a simple hole feature that does not have thread.

Tapped Hole. Click the Tapped Hole option if the hole is threaded. Thread information appears in the dialog box area so that you can specify the thread properties, as shown in the above image.

Taper Tapped Hole. Click the Taper Tapped Hole option if the hole is tapered thread. The taper tapped hole information appears in the dialog box area so you can specify the thread properties.

Center Points

Center points are sketched entities that can be used to locate hole features. To create a hole center, follow these steps:

1. Make a sketch active.
2. Click the Point, Center Point command in the Sketch tab > Draw panel as shown in the following image on the left.

3. Click a point to locate the hole center where you want to place the hole, and constrain and dimension the point as desired.

You can also switch between a hole center and a sketch point by using the Center Point command on the Sketch tab > Format panel, as shown in the following image on the right, when the sketch environment is active. When you press the button, you create a center point. If you use a center point style when you activate the From Sketch option of the Hole command, all center points that reside in the sketch are selected as centers for the hole feature automatically. This can expedite the process of creating multiple holes in a single hole feature. You can also select endpoints of lines, arcs, splines, center points of arcs and circles, or spline control points as hole centers. Points can be deselected by holding down the CTRL or SHIFT key and then selecting the points.

FIGURE 4.30

EXERCISE 4-2: CREATING HOLES

In this exercise, you will add drilled, tapped, and counterbored holes to a cylinder head.

1. Open *ESS_E04_02.ipt* in the Chapter 04 folder.
2. The first hole you place is a linear hole. Click the Hole command in the Modify panel.

   - For the Face, click the top planar face of Extrusion1.
   - For Reference 1 and 2, select the bottom left and left-angled edge, and place the point **.5 in** from each edge.
   - Verify that the Drilled hole option is selected.
   - Change the Termination to Through All.
   - Change the hole's diameter to **.25 in**, as shown in the following image.
   - Click OK to create the hole.

FIGURE 4.31

3. Next, you will create a hole based on the Sketch option. Click the Create 2D Sketch command on the Sketch panel, and then click the top planar face of Extrusion1. The center points of the arcs are projected onto the sketch.
4. Move the cursor to a blank area in the graphics window and right-click, and then click Finish Sketch from the menu.
5. Click the Hole command in the Modify panel or press the H key.

   - Click the center points that were projected from the two filleted edges.
   - Change the hole option to counterbore.
   - Ensure the Termination is set to Through All.
   - Change the hole's values as shown in the following image.
   - Click OK to create the holes.

FIGURE 4.32

6. Next, you create a tapped hole based on the Sketch option, but place it by first creating a center point. Click the Create 2D Sketch command and then click the top face of Extrusion1.
7. Click the View Face command from the Navigation bar, and click Sketch4 in the browser.
8. Click the Point, Center Point command in the 2D Sketch Panel, and place a point to the right of the angled edges.
9. Add a horizontal constraint between the left point where the angled edges converge.
10. Add a **.5 inch** dimension between the point that the angled edges converge and the center point as shown in the following image.

FIGURE 4.33

11. Finish the dimension command, right-click, and click Done.
12. Right-click in the graphics window, click Finish Sketch, the view should change to the home view. If not press the F6 key to change to the Home View.
13. Click the Hole command in the Modify panel bar or press the H key.
    - The Center Point is automatically selected as the Center.
    - Change the hole option to Drilled.
    - Change the hole type to Tapped.
    - Ensure the Termination is set to Through All.
    - Change the Thread Type to **ANSI Metric M Profile**.
    - Set the size to **.375** and the Designation is **3/8-16 UNC**, as shown in the following image.
    - Click OK to create the hole.

FIGURE 4.34

14. Next, you create a Taper Tapped hole that is concentric to the top of the cylinder.
    - Press the ENTER key to restart the Hole command.
    - Change the Placement option to Concentric.
    - For the Plane, select the top face of the cylinder.
    - For the Concentric Reference, select the top circular edge of the cylinder.
    - Ensure the hole option is set to Drilled.
    - Click the Taper Tapped Hole type.
    - Change the Thread Type to NPT.
    - Change the size to **1/2**.
    - Change the Termination to Through All, as shown in the following image.
    - Click Apply to create the hole.

FIGURE 4.35

15. For the last step, you create a tapped hole that uses the work point and work axis on the cylinder. Work point and axis will be covered later in this chapter.

    - Change the Placement option to On Point.
    - For the Point, select the work point on the cylinder.
    - For the Direction, select the work axis that goes through the cylinder.
    - Click the Tapped Hole type.
    - Change the Thread Type to Tapped Hole.
    - Change the size to **.25**.
    - Change the Termination to To and select the inside circular face of the tapered hole, as shown in the following image.
    - Click OK to create the hole.

FIGURE 4.36

16. The completed part is shown in the following image (the visibility of the work axis and work plane have been turned off for better visibility). Rotate the part and examine the holes.

FIGURE 4.37

17. Practice editing the holes and placing new holes.
18. Close the file. Do not save changes. End of exercise.

SHELLING

As you design parts, you may need to create a model that is made of thin walls. The easiest way to create a thin-walled part is to create the main shape and then use the Shell command to remove material.

The term *shell* refers to giving a wall thickness to the outside shape of a part and removing the remaining material. Essentially, you are scooping out the inside of a part and leaving the walls a specified thickness, as shown in the following image. You can offset the wall thickness in, out, or evenly in both directions. If the part you shell contains a void, such as a hole, the feature will have the thickness built around it.

A part may contain more than one shell feature, and individual faces of the part can have different thicknesses. If a wall has a different thickness than the shell thickness, it is referred to as a unique face thickness. If a face that you select for a unique face thickness has faces that are tangent to it, those faces will also have the same thickness. You can remove faces from being shelled, and these faces remain open. If no face is removed, the part is hollow on the inside.

FIGURE 4.38

To create a shell feature, follow these steps:

1. Create a part that will be shelled.
2. Issue the Shell command from the Model tab > Modify panel, as shown in the following image on the left.
3. The Shell dialog box appears, as shown in the following image on the right. Enter the data as needed.

FIGURE 4.39

4. After filling in the information in the dialog box, click OK, and the part is shelled.

To edit a shell feature, use one of the following methods:

- Right-click the name of the shell feature in the browser, and select Edit Feature from the menu. Alternately, you can double-click the feature's name or icon in the browser, and the Shell dialog box appears with all of the settings you used to create the feature. Change the settings as needed.
- Change the select priority to Feature Priority, double-click the shell feature, and the Shell dialog box appears with all of the settings you used to create the feature.

The following sections explain the options that are available for the Shell command.

Direction

Inside
Click this button to offset the wall thickness into the part by the given value.

Outside
Click this button to offset the wall thickness out of the part by the given value.

Both
Click this button to offset the wall thickness evenly into and out of the part by the given value.

Remove Faces
Click the Remove Faces button, and then click the face or faces to be left open. To deselect a face, click the Remove Faces button, and hold down the CTRL key while you click the face.

Automatic Face Chain
When you are removing faces and this option is checked, faces that are tangent to the selected face are automatically selected. Uncheck this option to select only the selected face.

Solids

If multiple solid bodies exist, select the solid body to shell.

Thickness

Enter a value or select a previously used value from the drop-down list to be used for the shell thickness.

Unique Face Thickness

Unique face thickness is available by clicking the More (>>) button that is located on the lower-right corner of the dialog box, as shown in the following image on the left.

To give a specific face a thickness, click on Click to add, select the face, and enter a value. A part may contain multiple faces that have a unique thickness, as shown in the following image on the right.

FIGURE 4.40

EXERCISE 4-3: SHELLING A PART

In this exercise, you will use the Shell command to create a shell on a part.

1. Open *ESS_E04_03.ipt* in the Chapter 04 folder.
2. Click the Shell command on the Modify panel.
3. Remove the top face by selecting the top face of the part.
4. Type a thickness of **.0625 in**, as shown in the following image.

FIGURE 4.41

5. Click OK to create the shell feature.
6. Rotate the part and examine the shell. The bottom face of the part should be closed.
7. Click the Home View.
8. Edit the Shell feature that you just created. Click on one of the inside faces of the shell and from the mini-toolbar click Edit Shell as shown in the following image.

FIGURE 4.42

9. Select the More (>>) button, click in the "Click to add" area, and select the left outside-vertical face.
10. Enter a value of **.5 in**, as shown in the following image.

FIGURE 4.43

11. Click OK to create the shell. Your part should resemble the following image.

FIGURE 4.44

12. Practice editing the shell feature; change the thickness direction, thickness and delete the unique thickness by selecting the unique thickness entry in the dialog box and then press the DELETE key.
13. Close the file without saving changes. End of exercise.

WORK FEATURES

When you create a parametric part, you define how the features of the part relate to one another; a change in one feature results in appropriate changes in all related features. Work features are special construction features that are attached parametrically to part geometry or other work features. You typically use work features to help you position and define new features in your model. There are three types of work features: work planes, work axes, and work points.

Use work features in the following situations:

- To position a sketch for new features when a planar part face is not available.
- To establish an intermediate position that is required to define other work features. You can create a work plane at an angle to an existing face, for example, and then create another work plane at an offset value from that plane.
- To establish a plane or edge from which you can place parametric dimensions and constraints.
- To provide an axis or point of rotation for revolved features and patterns.
- To provide an external feature termination plane off the part, such as a beveled extrusion edge, or an internal feature termination plane in cases where there are no existing surfaces.

Creating a Work Axis

A work axis is a feature that acts like a construction line. In the database, it is infinite in length but displayed a little larger than the model, and you can use it to help create work planes, work points, and subsequent part features. You can also use work axes as axes of rotation for polar arrays or to constrain parts in an assembly using assembly constraints. Their length always extends beyond the part—as the part changes size, the work axis also changes size. A work axis is tied parametrically to the part. As changes occur to the part, the work axis will maintain its relationship to the points, edge, or cylindrical face from which you created it. To create a work axis, use the Work Axis command on the Model tab > Work Features panel or press the key / (forward slash). The Axis command will allow you to select all geometry in the graphics window to create an axis. Another option is to click the down arrow in the lower right corner of the Axis button and select an option for creating an axis using only certain types of geometry as shown in the following image. These options will only select the predetermined geometry and will filter out all other types of geometry. For example, if the Through Two Points option is selected, only points can be selected in the graphics window.

You can also create a work axis when using the Work Plane or Work Point commands; right-click and select Create Axis when one of these work feature commands is active. Then, after the work axis is created, the command you were running will be active. The created axis will be indented as a child of the work axis or the work plane in the browser.

FIGURE 4.45

Use one of the following methods to create a work axis:

Axis Type	Selectable Geometry	Process
Axis	All geometry if available	Depending upon selected geometry.
On Line or Edge	Lines and linear edges	The work axis will be collinear with the selected line or linear edge.
Parallel to Line through Point	Linear edge or sketch line and an endpoint, midpoint, sketch point, or work point	The work axis will be parallel to the selected line or linear edge and through the selected point.
Through Two Points	Two endpoints, intersections, midpoints, sketch points, or work points	The work axis will go through the two selected points.
Intersection of Two Planes	Two nonparallel work planes or planar faces	The work axis will be created coincident with the intersection of the two planes.
Normal to Plane through Point	A planar face or work plane and an endpoint, midpoint, sketch point, or work point	The work axis is created perpendicular to the selected plane and through the point.
Through Center of Circular or Elliptical Edge	A circular edge, elliptical edge or an edge of a fillet	The work axis will create through the center point and perpendicular to the selected edge.
Through Revolved Face or Feature	A circular face	The work axis will create through the center of the selected circular face.

EXERCISE 4-4: CREATING WORK AXES

In this exercise, you will create a work axis to position a circular pattern. Circular patterns are covered later in this chapter.

1. Open *ESS_E04_04.ipt* in the Chapter 04 folder.
2. Click the Create 2D Sketch command and select the inside planar face as shown in the following image on the left.
3. If the view did not change, you are looking directly at the face, click the View Face command on the Navigation bar and then in the browser select the sketch you just created.
4. Click the Point, Center Point command in the Draw panel, and place a point near the middle of the sketch.
5. Place a vertical constraint between the center point you just created and the midpoint on the top line. Apply a horizontal constraint between the point and the midpoint of the edge on the right, as shown in the following image on the right.

FIGURE 4.46

6. Right-click in the graphics window, and click Done to finish the constraint command.
7. Right-click in the graphics window, and then select Finish Sketch.
8. If the view did not automatically change to the home view, right-click in the graphics window, and then click Home View from the menu.
9. From the Model tab > Work Features panel click the down arrow on the right of the Work Axis command and from the list click Normal to Plane through Point.
10. Select the angled planar face, and then select the center point. The work axis is created through this point and normal to the plane, as shown in the following image on the left.

NOTE One use of a work axis is to use it as a Rotation Axis when creating a circular pattern. The following image on the right shows a hole that was patterned around the work axis. Patterns are covered later in this chapter.

FIGURE 4.47

11. From the Model tab > Work Features panel click the down arrow on the right of the Work Axis command and from the list click Through Center of Circular or Elliptical Edge.
12. Select the top circular edge on the back of the part as shown in the following image on the left. The axis is created through the center of the circular edge.
13. Undo the last work axis you just created. Click the Undo command from the Quick Access toolbar and the work axis will removed.
14. From the Model tab > Work Features panel click the down arrow on the right of the Work Axis command and from the list click Through Revolved Face or Feature.
15. Select the back circular face on the part as shown in the following image on the right. The axis is created through the center of the circular face.

> Note that the results for the work axis are the same if the Through Center of Circular or Elliptical Edge or the option Through Revolved Face or Feature are used since both sets of geometry share the same centers.

NOTE

16. Practice creating work axes using different options.
17. Close the file. Do not save changes. End of exercise.

FIGURE 4.48

Creating Work Points

A work point is a feature that you can create on the active part or in 3D space and can be used to create other work features such as work axis and work planes, place a hole using the On Point option and place 3D lines on the work points. Work points are created relative to selected geometry, if the selected geometry changes location, the work point will move to keep the relationship to the new location of the geometry. To create a work point, use the Point command on the Model tab > Work Features panel, as shown in the following image, or press the key (.) (period). The point command will allow you to select all geometry in the graphics window to create a work point. Another option is to click the down arrow in the lower right corner of the Point button and select an option for creating a work point using only a filtered geometry type(s) as shown in the following image. These options will only select the predetermined geometry and will filter out all other types of geometry. For example, if the Intersection of Three Planes option is selected, only planes can be selected in the graphics window.

You can also create a work point when using the Work Plane or Work Axis commands; right-click and select Create Point when one of these work feature commands is active. Then, after the work point is created, the command you were running will be active. The created point will be indented as a child of the work axis or the work plane in the browser.

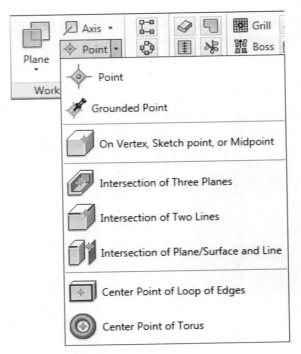

FIGURE 4.49

Use one of the following methods to create a work point:

Work Point Type	Process
Point	Select desired geometry, no filters are applied.
Grounded Point	Select a work point, midpoint, or vertex.
On Vertex, Sketch point, or Midpoint	Select a 2D or 3D sketch point, vertex, or the endpoint or midpoint of a line or linear edge.
Intersection of Three Planes	Select three work planes or planar faces.
Intersection of Two Lines	Select any two lines including linear edges, 2D or 3D sketch lines, or work axes.
Intersection of Plane/Surface and Line	Select a planar face or work plane and a work axis or line. Or select a surface and a sketch line, straight edge, or work axis.
Center Point of Loop of Edges	Select an edge of a closed loop of edges.
Center Point of Torus	Select a torus.

CREATING WORK PLANES

Before introducing work planes, it is important that you understand when you need to create a work plane. You can use a work plane when you need to create a sketch and no planar face exists at the desired location, or if you want a feature to terminate at a plane and no face exists to select. If you want to apply an assembly constraint to a plane on a part and no part face exists, you will need to create a work plane. If a face exists in any of these scenarios, you should use it and not create a work plane. A new sketch can be created on a work plane.

A work plane looks like a rectangular plane. It is tied parametrically to the part. Though extents of the plane will appear slightly larger than the part, the plane is in fact infinite. If the part or related feature moves or resizes, the work plane will also move or resize. For example, if a work plane is tangent to the outside face of a 10 mm diameter cylinder and the cylinder diameter changes to 20 mm, the work plane moves with the outside face of the cylinder. You can create as many work planes on a part as needed, and you can use any work plane to create a new sketch. A work plane is a feature and is modified like any other feature.

Before creating a work plane, ask yourself where this work plane needs to exist and what you know about its location. You might want a plane to be tangent to a given face and parallel to another plane, for example, or to go through the center of two arcs. Once you know what you want, select the appropriate options and create a work plane. There are times when you may need to create an intermediate work plane before creating the final work plane. You may need to create a work plane, for example, that is at 30° and tangent to a cylindrical face. You should first create a work plane that is at a 30° angle and located at the center of the cylinder; then create a work plane parallel to the angled work plane that is also tangent to the cylinder.

To create a work plane, click the Plane command on the Model tab > Work Features panel, or press the hot key (]) (end bracket). The Plane command allows you to select all geometry types in the graphics window to create a work plane. Another option is to click the down arrow in the lower right corner of the Plane button and select an option for creating a work plane using only a filtered geometry type as shown in the following image. These options will only select the predetermined geometry and will filter out all other types of geometry. For example if the Three Points option is selected, only points can be selected in the graphics window.

You can also create a work plane when using the Work axis or Work Point commands; right-click and select Create Plane when one of these work feature commands is active. Then, after the work plane is created, the command you were running will be active. The created plane will be indented as a child of the work axis or the work point in the browser.

Work planes can also be associated to the default reference (origin) planes, axis and points that exist in every part. These origin entities initially have their visibility turned off, but you can make them visible by expanding the Origin folder in the browser, right-clicking on a plane or planes, and selecting Visibility from the menu. You can use the origin planes to create a new sketch or to create other work planes.

FIGURE 4.50

Use one of the following methods to create a work plane:

Work Plane Type	Selectable Geometry	Process
Plane	Select vertices, edges, or faces	Click the desired geometry.
Offset from Plane	A planar face	Click a plane, drag the new work plane to a selected location. To specify the offset distance, enter a value in the input field.
Parallel to Plane through Point	A planar face, work plane and a point	Select a planar face or work plane and a point; order of selection does not matter.
Midplane between Two Parallel Planes	Planar faces or work planes	Create a work plane at the midplane defined by selecting two parallel planes.
Midplane of Torus	A torus	Create a work plane that goes through the center, or midplane, of a torus.
Angle to Plane around Edge	A face or plane and an edge or line parallel to the face	Create a work plane about an edge or line whose angle is determined from a planar face or plane. Enter the desired angle in the input field.
Three Points	Endpoints, intersections, midpoints, work points	Create a work plane based on three points.
Two Coplanar Edges	Work axes, edges, or lines	Select two coplanar work axes, edges, or lines.
Tangent to Surface through Edge	Circular face and a linear edge	Create a work plane that passes through the linear edge and is tangent to the circular face.
Tangent to Surface through Point	Circular face and an endpoint, midpoint, or work point	Create a work plane that passes through a point and is tangent to the circular face.
Tangent to Surface and Parallel to Plane	Circular face and a planar face or work plane, in either order	Create a work plane that is parallel to a plane and tangent to the circular face.
Normal to Axis through Point	Linear edge or axis and a point	Create a work plane that is perpendicular to an edge or axis and passes through a point.
Normal to Curve at Point	Nonlinear edge or sketch arc, circle, ellipse, or spline and a vertex, edge midpoint, sketch point, or work point on the curve	Create a work plane that is normal to a circular edge and passes through a point.

When you are creating a work plane, and more than one solution is possible, the Select Other command appears. Click the forward or reverse arrows from the Select Other command until you see the desired solution displayed. Click the checkmark in the selection box. If you clicked a midpoint on an edge, the resulting work plane links to the midpoint. If the selected edge's length changes, the location of the work plane will adjust to the new midpoint.

NOTE The order in which points or planes are selected is irrelevant.

Use the Show Me animations located in the Visual Syllabus to view animations that display how to create certain types of work planes.

UCS—User Coordinate System

A UCS is similar to the data in the origin folder; it contains three work planes, three axes, and a center point. Unlike the data in the origin folder you can create as many UCSs as required and position them as needed. The following list shows the common uses for a UCS:

- Create a UCS where it would be difficult to create a work plane, for example a compound angle.
- Locate a sketch on a UCS plane.
- Start and terminate features on UCS planes.
- A UCS axis can be used as a rotation axis to pattern features or parts.
- A UCS axis can be used to rotate parts.
- In an assembly, you can constrain UCSs of two parts.
- In a part or an assembly you can measure to the planes, axis or point in a UCS.
- Measure to the planes, axis or origin.

To start the UCS command, click Model tab > Work Features panel > UCS as shown in the following image.

FIGURE 4.51

There are multiple methods to place a UCS:

- In a part file, a UCS can also be placed on existing geometry.
- A UCS can be positioned using absolute coordinates in a part or an assembly file.

The first method is to place a UCS by selecting geometry:

- Start the UCS command, click Model tab > Work Features panel > UCS.
- Select a point to locate the origin.
- Select a point to define the direction of the X axis.
- Select a point to define the direction of the Y axis.

When selecting points to locate or position a UCS valid inputs are: vertex of an edge, midpoint of an edge, sketch, work point origin, solid circular edge, or solid elliptical edge.

The following image shows a UCS being placed at the top left vertex in the left image. The X axis is positioned by selecting a vertex as shown in the middle image, and the Y axis is also positioned by selecting a vertex as shown in the following image on the right. The midpoints of the edges could have also been used to align the UCS.

FIGURE 4.52

The second method to place a UCS is to enter absolute coordinates:

- Start the UCS command, click Model tab > Work Features panel > UCS.
- Enter a value for the X, Y, and Z location. Press the Tab key to switch between the cells as shown in the following image.
- After the three points are defined, press ENTER or click in the graphics window.
- To define the direction of the UCS, click on an arrow and enter a value.
- To rotate the UCS about an axis on the UCS, click on the shaft of an arrow and enter a value.

FIGURE 4.53

After creating a UCS, it will appear in the browser as shown in the following image. The UCS in the browser can be treated like the origin folder, turn the visibility on and off and measure to the planes, axis and center point.

FIGURE 4.54

To edit a UCS, move the cursor over the UCS in the graphics window or in the browser and double-click or right-click, and then click Redefine Feature as shown in the following image on the left. Select on the desired UCS segment: arrow, leg, or origin, and enter a new value or drag to a new location. When done relocating the UCS, right-click and click Finish from the menu. Do NOT click Done as this will cancel the operation. The following image on the right shows the prompt to select a segment to edit.

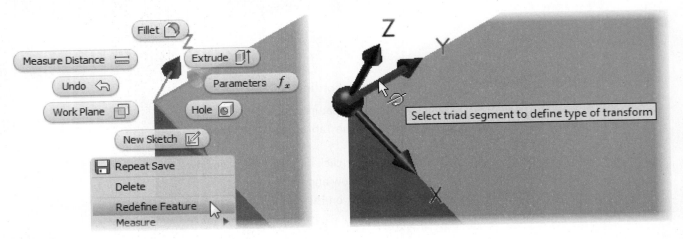

FIGURE 4.55

Feature Visibility

You can control the visibility of the origin planes, origin axes, origin point, user work planes, user work axes, user work points, and sketches by either right-clicking on them in the graphics area or on their name in the browser and selecting Visibility from the menu. You can also control the visibility for all origin planes, origin axes, origin point, user work planes, user work axes, user work points, sketches, solids, and UCS triad, planes, axis, and points from the View tab > Visibility panel > Object Visibility. Visibility can be checked to turn visibility on or cleared to turn it off, as shown in the following image.

FIGURE 4.56

EXERCISE 4-5: CREATING WORK PLANES AND A UCS

In this exercise, you will create work planes in order to create a boss on a cylinder head and a slot in a shaft.

1. Open *ESS_E04_05_1.ipt* in the Chapter 04 folder.
2. From the Model tab > Work Features panel, click the down arrow on the lower right corner of the Plane command and from the list click Angle to Plane around Edge.
3. Click the top rectangular face and the top-back left edge, and then enter a value of **-30** as shown in the following image on the left. Click the green checkmark in the mini-toolbar to create the work plane.
4. Make the top rectangular face the active sketch. Click on the top face of Extrusion1 and click Create Sketch command on the mini-toolbar.
5. Create and dimension a circle and apply a horizontal constraint between the center of the circle and the midpoint of the right vertical line as shown in the following image on the right. Finish the sketch.

FIGURE 4.57

6. Extrude the circle with the Extents set to To, and select the work plane, as shown in the following image.

FIGURE 4.58

7. In the browser, double-click on Work Plane1 the angled work plane you created in step 3 and enter different values for the angle; click the local Update command on the Quick Access toolbar to update the model. Change the angle of the work plane back to **–30 degrees**.

8. Turn off the visibility of the work plane by moving the cursor over an edge of the work plane in the graphics window, right-click, and click Visibility from the marking menu.

9. Another option to define a plane is to create a UCS. Click the UCS command on the Work Features panel.

10. Align the UCS on the top face of Extrusion1 by selecting the three vertices in the order as shown in the following image on the left.

11. Edit the UCS by double-clicking on one the arrowheads on the UCS in the graphics window.

12. Click the X leg of the UCS and enter a value of **–40** as shown in the following image on the right. To complete the edit, right-click and click Finish from the menu.

FIGURE 4.59

13. Reorder the UCS to it exists in the browser before Extrusion2. In the browser, drag the UCS entry so it is above Extrusion2 in the browser as shown in the following image. Reordering features is covered in Chapter 7.

FIGURE 4.60

14. Edit Extrusion2 and click the arrow button for the option Select surface to end the feature creation and in the browser, click UCS1: XY Plane in the UCS as shown in the following image.

FIGURE 4.61

15. Click OK to complete the edit.
16. Create a work plane that is centered between two parallel planes. From the Model tab > Work Features panel click the down arrow on the right side of the Plane command and from the list click Midplane between Two Parallel Planes.
17. Click the front-left face and the back-right face, as shown in the following image on the left.

18. Click one of the edges of the work plane you just created and click Create Sketch from the menu.
19. Slice the graphics by pressing the F7 key or right-click and click Slice Graphics from the menu.
20. To project the edges of the part, click the Project Cut Edges command from the Draw panel (the Project Cut Edges command may be under the Project Geometry command).
21. Create and dimension a circle on the top edge of the projected geometry, as shown in the following image on the right.

NOTE If the circle is placed at the midpoint of the projected edge, the horizontal dimension cannot be placed.

FIGURE 4.62

22. Finish the sketch, click Finish Sketch from the Exit panel.
23. Extrude the circle **.5** with the symmetric option, as shown in the following image.

FIGURE 4.63

24. Turn off the visibility of the work plane and the UCS. Your screen should resemble the following image on the left.

25. To verify that the symmetric extrusion will be maintained in the center of the part even if the width changes, edit Sketch1 of Extrusion1, and change the 1 inch dimension to **1.5 inches**.
26. If needed click the Local Update command on the Quick Access toolbar. Your screen should resemble the following image on the right.

FIGURE 4.64

27. Use the Free Orbit command to view the part from different perspectives. Notice how Extrusion2 still terminates at the UCS XY plane and Extrusion3 is still in the middle of the part.
28. Close the file. Do not save changes.
29. In this portion of the exercise, you will place two holes on a cylinder. Open *ESS_E04_05_2.ipt* from the Chapter 04 folder.
30. Create a work plane that is parallel to an origin plane and tangent to the cylinder.

    - In the browser, expand the Origin folder.
    - Create a work plane on the outside of the cylinder. From the Model tab > Work Features panel click the down arrow on the right side of the Work Plane command and from the list click Tangent to Surface and Parallel to Plane.
    - In the browser click the YZ plane under the Origin folder.
    - Click a point to the front right of the cylinder, as shown in the following image on the left.

31. Make the new work plane the active sketch.
32. Use the Project Geometry command from the Draw panel to project the Z axis of the origin folder onto the sketch.
33. Place a Point, Center Point on the projected axis—this will center the point in the center of the cylinder—and dimension it, as shown in the following image in the middle. Then finish the sketch.
34. Press H on the keyboard to start the Hole command and place a **.5 inch** through all hole at the center point.
35. Turn off the visibility of the work plane and your screen should resemble the following image on the right.

FIGURE 4.65

36. Next, you place a hole on the cylinder at an angle. First you create an angled work plane.

    - Create a work plane in the middle of the cylinder. From the Model tab > Work Features panel click the down arrow on the right side of the Plane command and from the list click Angle to Plane around Edge.
    - In the browser, click the YZ plane under the Origin folder.
    - In the browser click, the Z Axis under the Origin folder.
    - In the Angle dialog box, enter **315** as shown in the following image on the left.
    - Create the work plane, click the green check mark in the mini-toolbar.

37. Next, create a work plane that is parallel to an angled plane and tangent to the cylinder.

    - Create a work plane on the outside of the cylinder. From the Model tab > Work Features panel click the arrow on the right side of the Plane command and from the list click Tangent to Surface and Parallel to Plane.
    - Click the angle work plane that you just created.
    - Click a point to the front face of the cylinder, as shown in the following image on the right.

FIGURE 4.66

38. Turn off the visibility of the inside angled work planes. Move the cursor over the work plane. Right-click, and click Visibility from the menu.
39. Create a new sketch on the new work plane.
40. Use the Project Geometry command to project the Z axis of the origin folder onto the sketch.
41. Place a Point, Center Point on the projected axis, and dimension it, as shown in the following image on the left.
42. Finish the sketch.
43. Start the Hole command, and place a counterbore hole with the size of your choice at the center point.
44. Turn off the visibility of the work plane and your screen should resemble the following image on the right.

FIGURE 4.67

45. Expand Work Plane3 in the browser, double-click on Work Plane2, and enter a new value. Update the model to see the change.
46. Close the file. Do not save changes. End of exercise.

PATTERNS

There are two types of pattern methods: rectangular and circular. The pattern is represented as a single feature in the browser, but the original feature and individual feature occurrences are listed under the pattern feature. You can suppress the entire pattern or individual occurrences except for the first occurrence. Both rectangular and circular patterns have a child relationship to the parent feature(s) that you patterned. If the size of the parent feature changes, all of the child features will also change. If you patterned a hole, and the parent hole type changes, the child holes also change. Because a pattern is a feature, you can edit it like any other feature. You can pattern the base part or feature, as well as patterns. A rectangular pattern repeats the selected feature(s) along the direction set by two edges on the part or edges that reside in a sketch. These edges do not need to be horizontal or vertical, as shown in the following image. A circular pattern repeats the feature(s) around an axis, a cylindrical or conical face, or an edge.

FIGURE 4.68

Rectangular Patterns

When creating a rectangular pattern, you define two directions by clicking an edge or line segment in a sketch to define alignment. After you click the Rectangular Pattern on the Model tab > Pattern panel, as shown in the following image on the left, or press the hot key CTRL + SHIFT + R (rectangular pattern), the Rectangular Pattern dialog box appears, as shown in the following image on the right. Click one or more features to pattern, enter the values, and click the edges or axes as needed. A preview image of the pattern appears in the graphics area.

FIGURE 4.69

The following options are available for rectangular patterns.

Pattern Individual Features

Click this button to pattern a feature or features. When you select this option, you activate a features button, as described below.

Pattern the Entire Solid

Click this button to pattern a solid body. When you select this option, you select the entire part as the item to pattern. You also have the Include Work Features option when patterning an entire solid.

Features

Click this button, and then click a feature or features to be patterned from either the graphics window or the browser. You can add or remove features to or from the selection set by holding down the CTRL key and clicking them.

Solid. If multiple solid bodies exist, select the solid body that you want the feature(s) patterned on.

Direction 1

In Direction 1, you define the first direction for the alignment of the pattern. It can be an edge, an axis, or a path.

Path. Click this arrow button, and then click an edge or sketch that defines the alignment along which you will pattern the feature.

Flip. If the preview image shows the pattern going in the wrong direction, click this button to reverse its direction.

Midplane. Check this option to have the occurrences patterned on both sides of the selected feature. The midplane option is independent for both Direction 1 and Direction 2.

Column Count. Enter a value or click the arrow to choose a previously used value that represents the number of feature(s) you will include in the pattern along the selected direction or path.

Column Spacing. Enter a value or click the arrow to choose a previously used value that represents the distance between the patterned features.

Distance. Define the occurrences of the pattern using the provided dimension as the total overall distance for the patterned features.

Curve Length. Create the occurrences of the pattern at equal spacing along the length of the selected curve.

Direction 2

In Direction 2, you can define a second direction for the alignment of the pattern. It can be an edge, an axis, or a path, but it cannot be parallel to Direction 1.

Path. Click this button, and then click an edge or sketch that defines the alignment along which you will pattern the feature.

Flip. If the preview image shows the pattern going in the wrong direction, click this button to reverse its direction.

Midplane. Check this option to have the occurrences patterned on both sides of the selected feature. The midplane option is independent for both Direction 1 and Direction 2.

Row Count. Enter a value or click the arrow to choose a previously used value that represents the number of feature(s) that you will include in the pattern in the second direction.

Row Spacing. Enter a value or click the arrow to choose a previously used value that represents the distance between the patterned features.

Distance. Define the occurrences of the pattern using the provided dimension as the total overall distance for the patterned features.

Curve Length. Create the occurrences of the pattern at equal spacing along the length of the selected curve.

Options for the start point of the direction, compute type, and orientation method, as shown in the following image, are available by clicking the More (>>) button located in the bottom-right corner of the Rectangular Pattern dialog box.

FIGURE 4.70

Start

Click the Start button to specify where the start point for the first occurrence of the pattern will be placed. The pattern can begin at any selectable point on the part. You can select the start points for both Direction 1 and Direction 2.

Compute

Optimized. Click this option to pattern the feature's faces instead of the feature(s) to calculate all of the occurrences in the pattern. This option is ideal when the occurrences you are creating do not intersect and are all identical. It can improve the performance of pattern creation.

Identical. Click this option to use the same termination as that of the parent feature(s) for all of the occurrences in the pattern. This is the default option.

Adjust. Click this option to calculate the termination of each occurrence individually. Since each occurrence is calculated separately, the processing time can increase. You must use this option if a parent feature terminates to a face or plane.

Orientation

Identical. Click this option to orient all of the occurrences in the pattern the same as the parent feature(s). This is the default option.

Direction 1. Click this option to control the position of the patterned features by the selected direction. Each occurrence of the pattern is rotated to maintain proper orientation with the 2D tangent vector of the path.

Direction 2. Click this option to control the position of the patterned features by the selected direction. Each occurrence of the pattern is rotated to maintain proper orientation with the 2D tangent vector of the path.

Circular Patterns

When creating a circular pattern, you must have a work axis, a part edge, or a cylindrical face about which the features will rotate. After you click the Circular Pattern command on the Model tab > Pattern panel, as shown in the following image on the left, or press the hot key CTRL + SHIFT + O, the Circular Pattern dialog box appears, as shown in the following image on the right. Click a feature or features to pattern, enter the values, and click the edges and axis as needed. A preview image of the pattern appears in the graphics area.

FIGURE 4.71

The following options are available for circular patterns.

Pattern Individual Features
Click this button to pattern a feature or features. When you select this option, the features button is available and operates as described below.

Pattern the Entire Solid
Click this button to pattern a solid body. When you select this option, you select the entire part as the item to pattern. You also have the Include Work Features option when patterning an entire solid.

Features
Click this button, and then click a feature or features to be patterned. You can add or remove features to or from the selection set by holding down the CTRL key and clicking them.

Rotation Axis
Click the button and then click an edge, axis, or cylindrical face (center) that defines the axis about which the feature(s) will rotate.

Flip. If the preview image shows the pattern going in the wrong direction, click this button to reverse its direction.

Solid. If multiple solid bodies exist, select the solid body you want the feature(s) patterned on.

Placement

Occurrence Count. Enter a value or click the arrow to choose a previously used value that represents the number of feature(s) that you will include in the pattern. A positive number will pattern the feature(s) in the clockwise direction; a negative number will pattern the feature in the counterclockwise direction.

Occurrence Angle. Enter a value or click the arrow to choose a previously used value that represents the angle that you will use to calculate the spacing of the patterned features.

Midplane. Check this option to have the occurrences patterned evenly on both sides of the selected feature.

By clicking the More (>>) button, located in the bottom-right corner of the Circular Pattern dialog box, you can access options for the creation method and positioning method of the feature, as shown in the following image.

FIGURE 4.72

Creation Method

Optimized. Click this option to pattern the feature's faces instead of the feature(s) to calculate all of the occurrences in the pattern. This option is ideal when the occurrences you are creating do not intersect and are all identical. It can improve the performance of pattern creation. This method is recommended when there are 50 or more occurrences.

Identical. Click this option to use the same termination as that of the parent feature(s) for all of the occurrences in the pattern. This is the default option.

Adjust. Click this option to calculate each occurrence termination individually. Because each occurrence is calculated separately, the processing time can increase. You must use this option if a parent feature terminates to a face or plane.

Positioning Method

Incremental. Click this option to separate each occurrence by the number of degrees specified in Angle in the dialog box.

Fitted. Click this option to space each occurrence evenly within the angle specified in Angle in the dialog box.

> A work axis, an edge, or a cylindrical face (the centerline of the circular face will be used) about which the feature will rotate must exist before you create a circular pattern.

TIP

Linear Patterns—Pattern along a Path

You can also use the Rectangular Pattern command to pattern a feature about a path. You can define a path by a complete or partial ellipse, an open or closed spline, or a series of curves (lines, arcs, splines, etc.).

To pattern along a path, click the Path button, and use the options described above for rectangular patterns. The path you use can be either 2D or 3D.

EXERCISE 4-6: CREATING RECTANGULAR PATTERNS

In this exercise, you will create a rectangular pattern of holes to a cover plate.

1. Open *ESS_E04_06.ipt* in the Chapter 04 folder.
2. You pattern the hole. Click the Rectangular Pattern command in the Pattern panel.
3. Click the small hole feature as the feature to be patterned.
4. In the Direction 1 area of the dialog box, click the Path button, and then click the bottom horizontal edge of the part. A preview of the pattern is displayed.
5. Click the Flip button.
6. Enter **5** Count and **.625 in** Spacing.
7. Click the Direction 2 Path button, and click the vertical edge on the left side of the part.
8. Enter **4** Count and **.625 in** Spacing.

FIGURE 4.73

9. Click OK to create the pattern.

   You now suppress three of the holes that are not required in the design.

10. Expand the rectangular pattern feature entry in the browser to display the occurrences.

11. In the browser, move the cursor over the occurrences. Each occurrence highlights in the graphics window as you point to it in the browser.

12. Hold the CTRL key down, and click the three occurrences the holes will highlight on the model, as shown in the following image on the left.

13. Right-click on any one of the highlighted occurrences in the browser, and click Suppress, as shown in the following image.

FIGURE 4.74

14. The holes are suppressed in the model, as shown in the following image.

FIGURE 4.75

15. Practice editing the feature pattern, and change the count and spacing for each direction. The suppressed occurrences are still suppressed.

16. Close the file. Do not save changes. End of exercise.

EXERCISE 4-7: CREATING CIRCULAR PATTERNS

In this exercise, you will create a circular pattern of a counterbore hole.

1. Open *ESS_E04_07.ipt* in the Chapter 04 folder.
2. Edit the Hole1 feature to verify that the termination for the hole is Through All.
3. Click the Cancel button to close the Hole dialog box.
4. Click the Circular Pattern command in the Pattern panel.
5. Click the hole feature as the feature to pattern.
6. Click the Rotation Axis button in the Circular Pattern dialog box.
7. In the graphics window, click the work axis to specify the rotation axis. A preview of the pattern is displayed in the graphics window.
8. Type **8** in the Count field as shown in the following image.

FIGURE 4.76

9. Click the OK button to create the pattern.
10. Rotate the model to see the back side, as shown in the following image. The other six holes do not go through because the Identical Creation Method was selected; that is, the hole that is patterned is identical to the original hole. If desired, change the visual style to Wireframe to verify that all the holes are identical.

FIGURE 4.77

11. In the browser, right-click on Circular Pattern1 and click Edit Feature from the menu.
12. Click the More (>>) button.
13. Under Creation Method, select the Adjust option, as shown in the following image on the left.
14. Click OK to create the circular pattern. When done, your model should resemble the following image on the right.

FIGURE 4.78

15. Edit the circular pattern, and try different combinations of count, angle, creation method, and positioning method.
16. Close the file. Do not save changes. End of exercise.

EXERCISE 4-8: CREATING A PATTERN ALONG A NONLINEAR PATH

In this exercise, you will pattern a boss and hole along a nonlinear path.

1. Open *ESS_E04_08.ipt* in the Chapter 04 folder. Note that the visibility of the sketch that the pattern will follow is on.
2. Click the Rectangular Pattern command in the Pattern panel.
3. For the features to pattern, in the browser, click both Extrusion2 and Hole2 feature.
4. In the Direction 1 area in the Rectangular Pattern dialog box, click the Path arrow button. To select the entire path in the graphics window, click the sketch line near the extrusion feature. A preview of the pattern is displayed, as shown in the following image.

FIGURE 4.79

5. In Count, enter **40**, and in Spacing, enter **1.25**. The preview shows 1.25 inches between each occurrence.
6. From the Spacing drop-down list, select Distance. The preview shows 40 occurrences fit within 1.25 inches.
7. From the Distance drop-down list, select Curve Length. The preview shows the 40 occurrences fitting within the entire length of the path.
8. Rotate the model to verify that the occurrences on the right side are hanging off the model; this is because the first occurrence is spaced 1 inch away from the start of the path.
9. To solve the starting point issue, click the More (>>) button.
10. In the Direction 1 area in the dialog box, click the Start button, and then click the center point of the first hole, as shown in the following image on the left. The preview updates to show that all occurrences are now located on the part, as shown in the following image on the right. However, the last occurrence is on the edge of the part.

FIGURE 4.80

11. From the Curve Length drop-down list, select Distance. The curve length remains in the distance area but the value can be edited.
12. Click in the distance area and arrow to the right of the value of the four lines and three arcs. Subtract **–1 in** from the distance, as shown in the following image on the left.
13. Click OK to create the pattern.
14. In the browser, right-click on the entry Sketch, and uncheck Visibility from the menu. When done, your model should resemble the following image on the right.

FIGURE 4.81

15. Edit the pattern trying different combinations.
16. Close the file. Do not save changes.

APPLYING YOUR SKILLS

Skill Exercise 4-1

In this exercise, you will create a drain plate cover.

1. Start a new part based on the English *Standard (in).ipt* template.
2. Use the extrude, shell, hole, and rectangular pattern commands to create the part.

FIGURE 4.82

Skill Exercise 4-2

In this exercise, you will create a connector.

1. Start a new part based on the English *Standard (in).ipt* template.
2. Use the revolve, work plane, hole, chamfer, fillet, and circular pattern commands to complete the part.

FIGURE 4.83

CHECKING YOUR SKILLS

Use these questions to test your knowledge of the material covered in this chapter.

1. True ___ False ___ When creating a fillet feature that has more than one selection set, each selection set appears as an individual feature in the browser.
2. In regard to creating a fillet feature, what is a smooth radius transition?
3. True ___ False ___ When you are creating a fillet feature with the All Fillets option, material is removed from all concave edges.
4. True ___ False ___ When you are creating a chamfer feature with the Distance and Angle option, you can only chamfer one edge at a time.
5. True ___ False ___ When you are creating a hole feature, you do not need to have an active sketch.
6. What is a Point, Center Point used for?
7. True ___ False ___ A part may contain only one shell feature.
8. True ___ False ___ The only method to create a work axis is by clicking a cylindrical face.
9. True ___ False ___ You need to derive every new sketch from a work plane feature.
10. Explain the steps to create an offset work plane.
11. True ___ False ___ The options for creating work features filter out geometry that is not valid for the selected option.
12. True ___ False ___ You cannot create work planes from the default work planes.
13. True ___ False ___ A UCS can only be placed on existing geometry.
14. True ___ False ___ When you are creating a rectangular pattern, the directions along which the features are duplicated must be horizontal or vertical.
15. True ___ False ___ When you are creating a circular pattern, you can only use a work axis as the axis of rotation.
16. True ___ False ___ To start the Fillet command from the mini-toolbar, select the two planes that define the location of the fillet.
17. True ___ False ___ Use the Filet command with the Face option to create a fillet that is tangent to three faces.
18. While in the shell command, explain how to create a unique face thickness within the shell command.
19. Explain three reasons why you would create a UCS.
20. True ___ False ___ Use the Linear Pattern command to pattern a feature along a selected path, consisting of multiple lines and arcs.

CHAPTER 5

Creating and Editing Drawing Views

INTRODUCTION

After creating a part or assembly, the next step is to create 2D drawing views that represent that part or assembly. To create drawing views, start a new drawing file, select a 3D part or assembly on which to base the drawing views, insert or create a drawing sheet with a border and title block, project orthographic views from the part or assembly, and then add annotations to the views. You can create drawing views at any point after a part or assembly exists. The part or assembly does not need to be complete because the part and drawing views are associative in both directions (bidirectional). This means that if the part or assembly changes, the drawing views will automatically be updated. If a parametric dimension changes in a drawing view, the part will be updated before the drawing views get updated. This chapter will guide you through the steps for setting up styles, creating drawing views of a single part, editing dimensions, and adding annotations.

When creating drawings you adhere to specific drawing standards that communicate information about designs in a consistent manner. There are many different drawing standards, such as ANSI and ISO. Each company usually adopts a drawing standard and modifies it to meet their requirements. This book follows the ANSI drawing standard. More information about drawing standards and good practices are covered throughout this chapter.

OBJECTIVES

After completing this chapter, you will be able to perform the following:

- Make changes under the Drawing tab of the Application Options dialog box
- Create base and projected drawing views from a part
- Create auxiliary, section, detail, and broken views
- Edit the properties and location of drawing views
- Retrieve and arrange model dimensions for use in drawing views
- Edit, move, and hide dimensions
- Add automated centerlines

- Add general dimensions, baseline, chain, and ordinate dimensions
- Add annotations such as text, leaders, Geometric Dimensioning & Tolerancing (GD&T), surface finish symbols, weld symbols, and datum identifiers
- Create hole notes
- Open a model from a drawing
- Open a drawing from a model

Drawing Options

Before drawing views are created, the drawing options should be set to your preferences. To set the drawing options, click Application Options on the Tools panel. The Options dialog box will appear. Click the Drawing tab, your screen should resemble the following image. Make any changes to the options before creating the drawing views, otherwise the changes may not affect drawing views that you have already created. A description of a few of the options follow. For more information about the Drawing Application Options, consult the help system.

FIGURE 5.1

Center Dimension Text on Creation
Click this option to have dimension text centered as you create the dimension.

Edit Dimension When Created
Click this option to see the Edit Dimension dialog box everytime a dimension is placed in a drawing.

Default Drawing File Type
In this section, select either Inventor drawing IDW or DWG as the default file format. Also set how Inventor should handle DWG and drawing data.

Show Preview As. Sets how components are previewed when creating drawing views; All Components. Partial or Bounding Box. Partial and Bounding Box options reduce memory consumption. The preview has no effect on the computed drawing view.

Creating a Drawing

The first step in creating a drawing from an existing part or assembly is to create a new drawing *.idw* or *.dwg* file by one of the following methods. You can use the following methods to create a new part file:

- Click New on the Inventor Application menu, and then click the *Standard.idw* or *Standard.dwg* icon on the Default tab. You can also click an *.idw* or *.dwg* template on one of the other tabs.
- Click the New icon in the Quick Launch area of the Open dialog box, and then click the *Standard.idw* or *Standard.dwg* icon on the Default tab, as shown in the following image on the left, or click a *.idw* or *.dwg* template on one of the other tabs.
- From the Quick Access toolbar click the down arrow of the New icon and select Drawing from the menu as shown in the following image on the right. This creates a new drawing file based on the units that were selected when Autodesk Inventor was installed or on the *Standard.idw* or *Standard.dwg* that exists in the Templates folder.
- Use the shortcut key, CTRL-N, and then click the *Standard.idw or Standard.dwg* icon on the Default tab, as shown in the following image on the left. You can also click a *.idw* or *.dwg* template on one of the other tabs.

After starting a new drawing file using one of the previous methods, Autodesk Inventor's screen will change to reflect the drawing environment.

FIGURE 5.2

Seven drafting standard/templates, included with Autodesk Inventor, are presented under the Metric tab. If needed select one of these templates to start a drawing. They include the following:

- ANSI (American National Standards Institute)
- BSI (British Standards Institute)
- DIN (The German Institute for Standardization)
- GB (The Chinese National Standard)
- GOST (The Russian Standard)
- ISO (International Organization for Standardization)
- JIS (Japan Industrial Standard)

DWG TrueConnect

DWG TrueConnect allows you to work directly with AutoCAD *.dwg* files without the need for translations. This feature allows you to view, plot, and measure

AutoCAD drawing data while using Autodesk Inventor and perform the same actions to Autodesk Inventor drawing data while using AutoCAD, with the exception of editing the drawing views. The drawing view data in an Inventor .dwg file is controlled by Inventor. In this way, you can supply AutoCAD customers with .dwg drawings that contain familiar named items such as layers and object types such as blocks and title blocks. As was discussed in the Drawing Options section, from the Drawing tab of the Application Option dialog box set the Default Drawing File Type to .idw or .dwg. This sets the default file format that is used when a new drawing is created from the Quick Access toolbar > down arrow of the New icon > Drawing from the menu.

DRAWING SHEET PREPARATION

When you create a new drawing file using one of the provided template files or a template file that you created. Inventor displays a default drawing sheet with a default title block and border. The template .dwg or .idw file that is selected determines the default drawing sheet, title block, and border. The drawing sheet represents a blank piece of paper on which you can alter the border, title block, and drawing views. There is no limit to the number of sheets that can exist in the same drawing, but you must have at least one drawing sheet. To create a new sheet, click the New Sheet command from the Place Views tab > Sheets panel as shown in the following image on the left. Alternately, you can right-click in the browser and click New Sheet from the menu as shown in the following image in the middle, or on the current sheet in the graphics window right-click and click New Sheet from the marking menu as shown in the following image on the right.

FIGURE 5.3

A new sheet will appear in the browser, and the new sheet will appear in the graphics window with the same size, border and title block of the active sheet. To edit the sheet's size right-click on the sheet name in the browser, and select Edit Sheet from the menu, as shown in the following image on the left. Then select a size from the list, as shown in the following image on the right. To use your own values, select Custom Size from the list, and enter values for height and width. From the dialog box you can also change the sheet's name or slowly double-click on its name in the browser and then enter a new name.

FIGURE 5.4

 NOTE The sheet size is inserted full scale (1:1) and should be plotted at 1:1. The drawing views will be scaled to fit the sheet size.

Edit Property Fields Dialog Box

The time will come when you will need to fill in title block information. Expanding the default title block in the browser and double-click or right-click on the Field Text entry and click Edit Field Text from the menu as shown in the following image on the left will display the Edit Property Fields dialog box. By default, the following information will already be filled in: Sheet Number, Number of Sheets, Author, Creation Date, and Sheet Size. To fill in other title block information such as Part Number, Company Name, Checked By, and so on, select the iProperties command button as shown in the following image in the middle. The Drawing Properties dialog box will appear, as shown in the following image on the right, and you can fill in the information as needed. You can find most title block information under the Summary, Project, and Status tabs.

FIGURE 5.5

SAVE DRAWING DATA TO A TEMPLATE

After getting a drawing setup, you can save the drawing to the template folder so you can create a new drawing that contains the changes. From the Application Menu click Save As > Save Copy As Template as shown in the following image. Save the file to the correct folder; Templates (the default location), English and Metric or create a new folder. The location of the template files is set in the Application Options > File tab > Default templates or overridden in the active project's Folder Options > Templates.

FIGURE 5.6

CREATING DRAWING VIEWS

After you have set the drawing sheet format, border, and title block you can create drawing views from an existing part, assembly, or presentation file. The file from which you will create the views does not need to be open when a drawing view is created. It is suggested, however, that both the file and the associated drawing file be stored in the same directory and that the directory be referenced in the project file. When creating drawing views, you will find that there are many different types of views you can create. The following sections describe these view types.

Base View. This is the first drawing view of an existing part, assembly, or presentation file. It is typically used as a basis for generating the following dependent view types. You can create many base views in a given drawing file.

Projected View. This is a dependent orthographic or isometric view that is generated from an existing drawing view.

- Orthographic (Ortho View): A drawing view that is projected horizontally or vertically from another view.
- Iso View: A drawing view that is projected at a 30° angle from a given view. An isometric view can be projected to any of four quadrants.

Auxiliary View. This is a dependent drawing view that is perpendicular to a selected edge of another view.

Section View. This is a dependent drawing view that represents the area defined by a slicing plane or planes through a part or assembly.

Detail View. This is a dependent drawing view in which a selected area of an existing view will be generated at a specified scale.

Broken View. This is a dependent drawing view that shows a section of the part removed while the ends remain. Any dimension that spans over the break will reflect the actual object length.

Break Out View. This is a drawing view that has a defined area of material removed in order to expose internal parts or features.

Crop View. This drawing view allows a view to be clipped based on a defined boundary.

Overlay View. This drawing view uses positional representations to show an assembly in multiple positions in a single view.

> **NOTE** When you are creating drawing views Inventor will place a temporary raster image for each view while precise views are being calculated. While a precise view is being calculated a small horizontal and a vertical line will appear in the corners of the view and two circular arrows are displayed in the browser in front of the views that are being calculated. While the view(s) are being calculated you can continue to work on the drawing.

CREATING A BASE VIEW

A base view is the first view that you create from the selected part, assembly, or presentation file. When you create a base view, the scale is set in the dialog box, and from this view, you can project other drawing views. There is no limit to the number of base views you can create in a drawing based on different parts, assemblies, or presentation files. As you create a base view, you can select the orientation of that view from the Orientation list on the Component tab of the Drawing View dialog box. By default there is an option to create projected views immediately after placing a base view.

To create a base view, follow these steps:

1. Click the Base View command on the Place Views tab as shown in the following image on the left, or right-click in the graphics window and select Base View.
2. The Drawing View dialog box will appear, also shown in the following image. On the Component tab, click the Open an existing file icon to navigate to and select the part, assembly, or presentation file from which to create the base drawing view. After making the selection, a preview image will appear attached to your cursor in the graphics window. Do not place the view until the desired view options have been set.
3. Select the type of view to generate from the Orientation list.
4. Select the scale for the view.
5. Select the style for the view.
6. By default the option is on that allows you to create projected views immediately after placing a base view. Uncheck this option if you want to only place a base view.
7. Locate the view by selecting a point in the graphics window.

As shown in the following image on the right, the Drawing View dialog box has three tabs: Component, Model State, and Display Options. The following sections describe these tabs.

FIGURE 5.7

The Component Tab

The following categories are explained in detail under the Drawing View Component tab.

File. Any open part, assembly, or presentation files will appear in the drop-down list. You can also click the Explore directories icon and navigate to and select a part, assembly, or presentation file.

Orientation. After selecting the file, choose the orientation in which to create the view. After selecting an orientation, a preview image will appear in the graphics window attached to your cursor. If the preview image does not show the view orientation that you want, select a different orientation view by selecting another option. A symbol that identifies the type of drawing projection is also displayed under the Orientation area and is a symbol that identifies the type of drawing projection, such as First Angle or Third Angle Projection.

The following view orientations are available:

- Front, Current (current orientation of the part, assembly, or presentation in its file), Top, Bottom, Left, Right, Back, Iso Top Right, Iso Top Left, Iso Bottom Right, and Iso Bottom Left. A Change view orientation button is also present that allows you to create a custom view.

View/Scale Label (Light Bulb). Click to display or hide the view scale label. Clicking the checkbox displays the scale label, and leaving the box unchecked hides the scale label.

Scale. Enter a number for the scale in which to create the view. Note that the drawing sheet will be plotted at full scale (1:1), and the drawing views are scaled as needed to fit the sheet. You can edit the scale of the views after the views have been generated.

Scale from Base. Click to set the scale of a dependent view to match that of its base view, as shown in the following image on the left. To change the scale of a dependent view, clear the option's box, and modify the scale value for that view. This feature activates when you edit existing views.

View Identifier. Use to include and/or change the label for the selected view. When you create a view, a default label is determined by the active drafting standard. To change the label, select the label in the box and enter the new label.

Style. Choose how the view will appear. There are three choices: Hidden Line, Hidden Line Removed, and Shaded. The preview image will not update to reflect the style choice. When the view is created, the chosen style will be applied. The style can be edited after the view has been generated.

Style from Base. Click to set the display style of a dependent view to be the same as that of its base view, as shown in the following image on the right. To change the display style of a dependent view, clear the checkbox and modify the style for that view. This feature activates when you edit existing views.

Feature Preview. Click to preview the drawing view before it is created. This option is unavailable if Show Preview As option is set to All Components on the Drawing tab of the Application Options dialog box.

Create Projected Views Immediately after Base View Creation. By default the option is on that allows you to create projected views immediately after placing a base view. Uncheck this option if you want to only place a base view.

The Model State Tab

Use this tab to specify the weldment state and member states of an iAssembly or iPart to use in a drawing view, as shown in the image below. Other items are used to control line style and hidden line calculations.

Weldment. Enable this area when you select a document that is a weldment. Weldments have four states: Assembly, Preparations, Welds, and Machining.

Member. Allows you to select the member of an iAssembly or iPart to represent in a drawing view.

Line Style. Three options are presented: As Reference Parts, As Parts, and Off. When you select As Reference Parts for shaded view types, the reference data will appear transparently shaded with tangent edges turned off and all part edges as phantom lines. The reference data will appear on top of the product data that is shaded. When you select As Parts, reference data will have no special display characteristics. The reference data will appear on top of product data, and the reference data will appear with tangent edges turned off and all other edges as the phantom line type. When you select Off, reference data will not appear.

Hidden Line Calculation. Specifies if hidden lines are calculated for Reference Data Separately or for All Bodies.

Margin. To see more reference data, set the value to expand the view boundaries by a specified value on all sides.

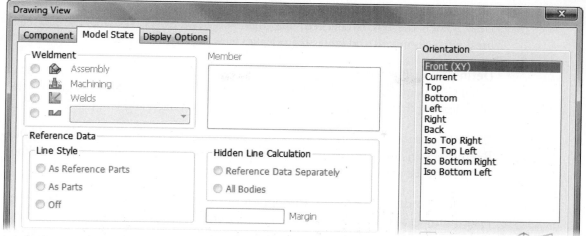

FIGURE 5.8

The Display Options Tab

The following image shows the Display Options tab of the Drawing View dialog box. The following sections describe the tab's options.

All Model Dimensions. Click to see model dimensions in the view, which is only active upon base view creation. If the option's box is clear, model dimensions will not be placed automatically upon view creation. When checked, only the dimensions that are parallel to the view and have not been retrieved in existing views on the sheet will appear.

Model Welding Symbols. This box is active only if you are creating a drawing view of a weldment. Click to retrieve welding symbols placed in the model in the drawing.

Bend Extents. This box is active only if you are creating a drawing view of a sheet metal part flat pattern. Click to control the visibility of bend extent lines or edges.

Thread Feature. This box is active only if you are creating a view of an assembly model. Click to set the visibility of thread features in the view.

Weld Annotations. This box is active only if creating a view of a weldment. Click to control the display of weld annotations.

User Work Features. Click to display work features in the drawing view.

Interference Edges. When selected, a drawing view will display both hidden and visible edges due to an interference condition such as a press or interference fit.

Tangent Edges. Click to set the visibility of tangent edges in a selected view. Checking the box displays tangent edges, and leaving the box clear hides them. If enabled, tangent edges can also be adjusted to display Foreshortened.

Show Trails. Click to control the display of trails in drawing views based on presentation files.

Hatching. Click to set the visibility of the hatch lines in the selected section view. Checking the box displays hatch lines, and leaving the box clear hides them.

Align to Base. Click to remove the alignment constraint of a selected view to its base view. When the box is checked, alignment of views exists. Leaving the box clear breaks the alignment and labels the selected view and the base view.

Definition in Base View. Click to display or hide the section view projection line or detail view boundary. Checking the box displays the line, circle, or rectangle and text label; leaving the box clear hides the line, circle, or rectangle and text label.

Cut Inheritance. Use this option to turn on and off the inheritance of a Break Out, Break, Section, and Slice cut for the edited view. Selecting the appropriate checkbox will inherit the corresponding cut from the parent view.

Section Standard Parts. Use this option to control whether or not standard parts placed into an assembly from the supplied Inventor parts library, such as nuts, bolts, and washers, are sectioned. Three options are available in this area: Always, Never, and Obey Browser Settings.

View Justification. Use to control the position of drawing views when the size or position of the model changes. This area is especially helpful with creating drawing views from assembly models. This area contains two modes: Centered and Fixed. The Centered mode keeps the model image centered in the drawing view. If, however, the model's view increases or decreases in size, the drawing view will shift on the drawing sheet. In some cases, this shift could overlap the drawing border or title block. The Fixed mode keeps the drawing view anchored on the drawing sheet. In the event that the model image increases, an edge of the drawing view will remain fixed.

FIGURE 5.9

Creating Projected Views

A projected view can be an orthographic or isometric view that you project from a base view or any other existing view. When you create a projected view, a preview image will appear, showing the orientation of the view you will create as the cursor moves to a location on the drawing. There is no limit to the number of projected views you can create. To create a projected drawing view, follow these steps:

1. Click the Projected command on the Place Views tab > Create panel as shown in the following image. You could also right-click inside the bounding area of an existing view box, displayed as dashed lines when the cursor moves into the view. Select Create View > Projected View from the menu.

FIGURE 5.10

2. If you selected the Projected command, click inside the desired view to start the projection.
3. Move the cursor horizontally, vertically, or at an angle (to create an isometric view) to get a preview image of the view you will generate. Keep moving the cursor until the preview matches the view that you want to create, and then press the left mouse button. Continue placing projected views.
4. When finished, right-click, and select Create from the menu.

EXERCISE 5-1: CREATING A MULTIVIEW DRAWING

In this exercise, you will create a base view, and then you will add projected views to create a multiview orthographic drawing. Finally, you will add an isometric view to the drawing.

1. Open the file *ESS_E05_01.idw*. in the Chapter 05 folder. This drawing file contains a single sheet with a border and title block. In addition to the title block information, notice also the numbers and letters present on the outside of the border. Use these as reference points to locate a specific detail, view, or dimension on a large sheet. Numbers are located horizontally along the top and bottom of the border. Letters are located vertically along the left and right border edges. A typical references example would be C2. Here, you would identify the C along the vertical portion of the border and the number 2 along the horizontal portion of the border. Where both of these two references intersect, your search item should be easily found.
2. Create a base view by clicking on the Base View command; this will display the Drawing View dialog box.
3. Under the File area of the Component tab, click the Open an existing file button (1), and double-click the file *ESS_E05_01.ipt*. in the Chapter 05 folder to use it as the view source.
4. In the Orientation area, verify that Front (XY) is selected (2).
5. In the Scale list, verify that 1:1 is current (3).

6. Verify that the Create projected views immediately after base view creation option is checked (4).
7. In Style, verify that the Hidden Line button in the lower right corner of the Drawing View dialog box is selected (5). The following image shows all the settings in the Drawing View dialog box.

FIGURE 5.11

8. Click the Display Options tab, and ensure that All Model Dimensions is not checked.
9. Position the view preview in the lower-left corner of the sheet (in Zone C6), and then click to place the view as shown in the following image.

FIGURE 5.10

10. Next you place projected views. While still in the Base View command move the cursor to the right of the base view. Click in Zone C2 to place the right-side view.
11. Move the cursor to the top of the base view. Click in Zone E6 to place the top view.
12. Right-click, and click Create from the menu to create the views, as shown in the following image.

> **NOTE** You could have created the right side and top view using the Projected view command.

FIGURE 5.13

13. The views are crowded with a scale of 1:1. In the steps that follow, you will reduce the size of all drawing views by setting the base view scale to 1:2. The dependent views will update automatically. Edit the base view (front view) by either double-clicking in the base view or right-click in the base view and click Edit View from the menu.

> **NOTE** To edit the view, right-click on the view border or inside the view. Do not right-click on the geometry.

14. When the Drawing View dialog box displays, select 1 / 2 from the Scale list, and click OK.
15. The scale of all views updates, as shown in the following image.

FIGURE 5.14

16. Now change the drawing scale in the title block. Begin this process by expanding Sheet:1 in the browser.
17. From the browser, also expand the title block ANSI-Large under Sheet:1.
18. Right-click on the Field Text icon, and choose Edit Field Text from the menu.
19. The Edit Property Fields dialog box will display. In the SCALE cell, click on 1:1 and enter a new scale of 1:2 as shown in the following image on the left.
20. Click OK. The title block updates to the new scale, as shown in the following image on the right.

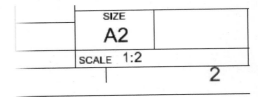

FIGURE 5.15

21. To complete the drawing views, an isometric view will be created. Begin by clicking the Projected view command on the Place Views tab > Create panel.
22. Select the base view (front view) and move your cursor to a point above and to the right of the base view.
23. Click in Zone E3 to place the isometric view. Right-click the sheet and choose Create. Your drawing should appear similar to the following image.

FIGURE 5.16

24. Double-click the isometric view and click the Display Options tab.
25. In the Display tab, clear the check box for Tangent Edges, and then click OK.
26. Complete this exercise by moving drawing views to better locations. Select the right-side view, and drag it to Zone C4. Select the isometric view, and drag it to Zone E4. Your drawing should appear similar to the following image.

FIGURE 5.17

27. Close all open files. Do not save changes. End of exercise.

Creating Auxiliary Views

An auxiliary view is a view that is projected perpendicular to a selected edge or line in a base view. It is designed primarily to view the true size and shape of a surface that appears foreshortened in other views.

To create an auxiliary drawing view, follow these steps:

1. Click the Auxiliary command on the Place Views tab, as shown in the following image. You can also right-click inside the bounding area of an existing view box, shown as a dotted box when the cursor moves into the view, and then select Create View > Auxiliary View from the menu.

FIGURE 5.18

2. If you selected the Auxiliary command, click inside the view from which the auxiliary view will be projected. The Auxiliary View dialog box will appear, as shown in the following image on the left. Type in a name for Label and a value for Scale, and then select one of the Style options: Hidden, Hidden Line Removed, or Shaded.
3. In the selected drawing view, select a linear edge or line from which the auxiliary view will be perpendicularly projected, as shown in the following image on the right.

FIGURE 5.19

4. Move the cursor to position the auxiliary view, as shown in the following image on the left. Notice that the view takes on a shaded appearance as you position it.
5. Click a point on the drawing sheet to create the auxiliary view. The completed auxiliary view layout is shown in the following image on the right.

FIGURE 5.20

Creating Section Views

A section view is a view you create by defining a section line or multiple section lines that will represent the plane(s) that will be cut through a part or assembly. The view will represent the surface of the cut area and any geometry shown behind the cut face from the direction being viewed. When defining a section, you sketch line segments that are horizontal, vertical, or at an angle. You cannot use arcs, splines, or circles to define section lines. When you sketch the section line(s), geometric constraints will automatically be applied between the line being sketched and the geometry in the drawing view. You can also infer points by moving the cursor over (or scrubbing) certain geometry locations, such as centers of arcs, endpoints of lines, and so on, and then moving the cursor away to display a dotted line showing that you are inferring or tracking that point. To place a geometric constraint between the drawing view geometry and the section line, click in the drawing when a green circle appears; the glyph for the constraint will appear. If you do not want the section lines to have constraint(s) applied to them automatically when they are created, hold down the CTRL key when sketching the line(s). Because the area in the section view that is solid material appears with a hatch pattern by default, you may want to set the hatching style before creating a section view.

To create a section drawing view, follow these steps:

1. Click the Section command on the Place Views tab > Create panel, as shown in the following image. You can also right-click inside the bounding area of an existing view, displayed as a dotted box when the cursor moves into the view. You can then select Create View > Section from the menu.
2. If you selected the Section command, click inside the view from which to create the section view.
3. Sketch the line or lines that define where and how you want the view to be cut. In the following image, a vertical line is sketched through the center of the object. When sketching and a green dot appears a coincident constraint will be applied.

FIGURE 5.21

4. When you finish sketching the section line(s), right-click and select Continue from the menu as shown in the following image on the left.
5. The Section View dialog box will appear, as shown in the following image on the right. Fill in the information for how you want the label, scale, and style to appear in the drawing view. When the Include Slice option is checked, a section view will be created with some components sliced and some components sectioned, depending on their browser attribute settings. Placing a check in the box next to Slice All parts will override any browser component settings and will slice all parts in the view according to the Section line geometry. Components that are not crossed by the Section Line will not be included in the slicing operation.

FIGURE 5.22

6. Move the cursor to position the section view (the section arrows will flip direction depending upon the location of the cursor), as shown in the following image, and select a point to place the view.

FIGURE 5.23

Creating Aligned Sections

Aligned sections take into consideration the angular position of details or feature of a drawing instead of projecting the section view perpendicular to the view, as shown in the following image on the left. As illustrated on the left in the following image, it is difficult to obtain the true size of the angled elements of the bottom area in the section view. In the Side view, they appear foreshortened or not to scale. Hidden lines were added as an attempt to better clarify the view.

An aligned section view creates a section view that is perpendicular to its section lines. This prevents objects in the section view from being distorted.

PROJECTED   **ALIGNED**

FIGURE 5.24

To create an aligned section view, follow these steps:

1. Click the Section command on the Place Views tab. You can also right-click inside the bounding area of an existing view box, shown as a dotted box when the cursor moves into the view. You can then select Create View > Section from the menu.
2. If you selected the Section command, click inside the view from which to create the aligned section view.
3. Sketch the line or lines that define where and how you want the view to be cut. In the following image, a vertical line is sketched up to the center of the object. Then, the line changes direction to catch the counterbore feature as shown in the following image on the left. When sketching and a green dot appears a coincident constraint will be applied.
4. When you finish sketching the section line(s), right-click and select Continue.
5. The Section View dialog box will appear, as shown in the following image on the right. Fill in the information for how you want the label, scale, and style to appear in the drawing view. Verify that the type of section being created is Aligned. When the Include Slice option is checked, a section view will be created with some components sliced and some components sectioned, depending on their browser attribute settings. Placing a check in the box next to Slice All parts will override any browser component settings and will slice all parts in the view according to the Section line geometry. Components that are not crossed by the Section Line will not be included in the slicing operation.

FIGURE 5.25

6. The completed aligned section is displayed in the following image. The bottom counterbore hole was rotated until it appears as a full section.

7. To edit the location of the section line(s), right-click on the section line and click Edit from the menu. The sketch environment will be current, add or delete constraints as needed. Dimensions can also be added between a section line and existing geometry. Before adding dimensions use the Project Geometry command to project the existing geometry onto the active sketch.

FIGURE 5.26

Modifying Section View and a Hatch Pattern

You can edit the section view by right-clicking in the section view and clicking Edit Section Properties. In the Edit Section dialog box you can change the section depth, to include a slice and change the method to Projected or Aligned if the view supports it. You can also edit the hatch pattern by right-clicking on the hatch pattern in the section view and selecting Edit from the menu. This will launch the Edit Hatch Pattern dialog box, as shown in the following image. Make the desired changes in the

Pattern, Angle, Scale, Line Weight, Shift (shifts the hatch pattern a specified distance, but stills stays within the section boundary), Double, and Color areas. Click the OK button to reflect the changes in the drawing.

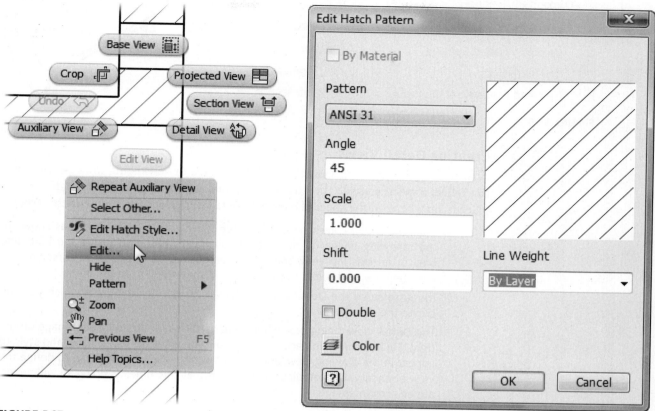

FIGURE 5.27

Hatching Isometric Views

In addition to applying crosshatching to orthographic view, as shown in the following image on the left, you can also apply crosshatching to an isometric view. To perform this operation, double-click or right-click inside of the bounding box that holds the isometric view, and pick Edit View from the menu, as shown in the following image.

FIGURE 5.28

When the Drawing View dialog box appears, click on the Display Options tab, and place a check in the box next to Hatching, as shown in the following image on the left. The results are illustrated in the following image on the right with the hatching applied to the isometric view.

FIGURE 5.29

Creating Detail Views

A detail view is a drawing view that isolates an area of an existing drawing view and can reflect a specified scale. You define a detailed area by a circle or rectangle and can place it anywhere on the sheet. To create a detail drawing view, follow these steps:

1. Click the Detail command on the Place Views tab > Create panel as shown in the following image on the left. You can also right-click inside the bounding area of an existing view box, shown as dashed lines when the cursor moves into the view, and select Create View > Detail View from the menu.
2. If you selected the Detail command, click inside the view from which you will create the detail view.
3. The Detail View dialog box will appear, as shown in the following image on the right. Fill in information according to how you want the label, scale, and style to appear in the drawing view, and pick the desired view fence shape to define the view boundary. Do not click the OK button at this time, as doing so will generate an error message.

FIGURE 5.30

4. In the selected view, select a point to use as the center of the fence shape that will describe the detail area. The following image on the left shows the icon that appears when creating a circular fence. The image on the right shows the icon used for creating a rectangular fence.

FIGURE 5.31

> **NOTE** You can right-click at this point and select Rectangular Fence to change the bounding area of the detail area definition.

5. Select another point that will define the radius of the detail circle, as shown in the following image on the left, or corner of the rectangle, as shown in the following image on the right. As you move the cursor, a preview of the boundary will appear.

FIGURE 5.32

6. Select a point on the sheet where you want to place the view. There are no restrictions on where you can place the view. The following image on the left shows the completed detail view based on a circular fence; a similar detail based on a rectangular fence is shown on the right.

FIGURE 5.33

Modifying a Detail View

After a detail view is created, numerous options are available to fine-tune the detail. To access these controls, right-click on the edge of the detail circle and choose options, as shown in the following image on the right. The three detail view options are explained below.

- **Smooth Cutout Shape:** Placing a check in this box will affect actual detail view. In the following image, instead of the cutout being ragged, a smooth cutout shape is present.
- **Full Detail Boundary:** This option displays a full detail boundary in the detail view when checked, as shown in the following image.
- **Connection Line:** This option will create a connection line between the main drawing and the detail view, as shown in the following image.

FIGURE 5.34

NOTE Once a connection line has been created in a detail view, you can add a new vertex by right-clicking the detail boundary or connection line and clicking Add Vertex from the menu. Picking a point on the connection line will allow you to add the vertex and then move the new vertex to the new position. You can even remove a vertex by right-clicking a vertex and choosing Delete Vertex from the menu.

Moving Drawing Views

To move a drawing view, move the cursor over the view until a bounding box consisting of dotted lines appears, as shown in the following image. Press and hold down the left mouse button, and move the view to its new location. Release the mouse button when you are finished. As you move the view, a rectangle will appear that represents the bounding box of the drawing view. If you move a base view, any projected children or dependent views will also move with it as required to maintain view alignment. If you move an orthographic or auxiliary view, you will only be able move it along the axis in which it was projected from the part edge or face. You can move detail and isometric views anywhere in the drawing sheet.

FIGURE 5.35

EXERCISE 5-2: CREATING AUXILIARY, SECTION, AND DETAIL VIEWS

In this exercise, you create a variety of drawing views from a model of a cover.

1. Open the file *ESS_E05_02.idw*. Use the Zoom command to zoom in on the top view.
2. First you create an auxiliary view. Click the Auxiliary command on the Place Views tab > Create panel.
3. In the graphics window, click in the top view.
4. Select the left-outside angled line, as shown in the following image on the left, to define the projection direction.
5. Place the view to the left as shown in the following image on the right.

FIGURE 5.36

6. In the graphics window, select and drag the border of the views to fit them in the drawing.
7. You now create a section view. Click the Section command on the Place Views tab > Create panel.
8. In the graphics window, click in the top view.
9. To define the first point of the section line, hover the cursor over the center hole, and then click a point directly above the hole (the dotted lines represent the inferred point), as shown in the following image.
10. Click a point below the geometry to create a vertical section line.
11. Right-click and click Continue from the menu.
12. Place the section view with the default settings to the right of the top view, as shown in the following image on the right.
13. If the section view does not show the entire part, drag the endpoints of the section line in the top view so it is above the view.

FIGURE 5.37

14. Try to drag the section line in the top view. If the section line can move follow these steps.
    - Right-click on the section line and click Edit from the menu.
    - Use the project geometry command and project the circle that is in the center of the part.
    - Apply a coincident constraint between the section line and the center point of the projected circle.
    - Click the Finish Sketch command from the Exit panel.

> **NOTE** If the sketch line cannot move and you need to reposition it, edit the section line as described in step 14, but remove a constraint that is holding it in place.

15. If desired, project an isometric view from the sectioned view.
16. Delete the section view by right-clicking in the section view and click Delete from the menu.
17. Next you create an aligned section view. Click the Section command on the Place Views tab > Create panel.
18. In the graphics window, click in the top view.
19. To define the first point of the section line, hover the cursor over the center hole, and then click a point directly above the hole (the dotted lines represent the inferred point), as you did in step 9.
20. Click the center point of the center circle.
21. Move the cursor over the center point of the lower-right circle (a green dot will appear) and then click a point outside of the geometry as shown in the following image on the left.
22. Right-click and click Continue from the menu.
23. Place the section view with the default settings to the right of the top view, as shown in the following image on the right.

FIGURE 5.38

24. If the section view does not show the entire part, drag the endpoints of the section line in the top view so the entire part is sectioned.
25. Finally, create a detail view. Click the Detail command from the Place Views tab > Create panel.
26. In the graphics window, click in the top view.
27. Select a point near the left corner of the view to define the center of the circular boundary of the detail view.
28. Drag the circular boundary to include the entire lower-left corner of the view geometry, as shown in the following image on the left.
29. Click to position the detail view below the break view. The following image in the middle shows the completed detail in the top view and the image on the right shows the completed detail view.

Your view identifier (letter) may be different than what is shown.   **NOTE**

FIGURE 5.39

30. Reposition the drawing views as shown in the following image.

FIGURE 5.40

31. Close all open files. Do not save changes. End of exercise.

Creating Break Views

When creating drawing views of long parts, you may want to remove a section or multiple sections from the middle of the part in a drawing view and show just the ends. This type of view is referred to as a "break view." You may, for example, want to create a drawing view of a 2 × 2 × 1/4 angle, as shown in the following image, which is 48 inches long and has only the ends chamfered. When you create a drawing view, the detail of the ends is small and difficult to see because the part is so long.

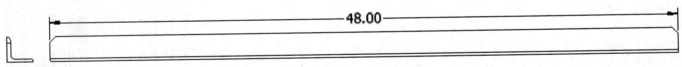

FIGURE 5.41

In this case, you can create a break view that removes the middle of the angle and leaves a small section on each end. When you place an overall length dimension that spans the break, it appears as 48 inches, and the dimension line shows a break symbol to note that it was based from a break view, as shown in the following image.

FIGURE 5.42

You create a break view by adding breaks to an existing drawing view. The view types that can be changed into break views are as follows: part views, assembly views, projected views, isometric views, auxiliary views, section views, break out views, and detail views. After creating a break view, you can move the breaks dynamically to change what you see in the broken view.

To create a break view, follow these steps:

1. Create a base or projected view or one that will eventually be shown as broken.
2. Click the Break command on the Place Views tab > Modify panel, as shown in the following image on the left. You can also right-click inside the bounding area of an existing view box, shown as a dotted rectangle when the cursor is moved into the view. Select Create View > Break from the menu.
3. If you selected the Break command on the Place Views tab > Modify panel, click inside the view from which to create the break view.
4. The Break dialog box will appear, as shown in the following image on the right. Do not click the OK button at this time, as this will end the command.

FIGURE 5.43

5. In the drawing view that will be broken, select a point where the break will begin, as shown in the following image.

FIGURE 5.44

6. Select a second point to locate the second break, as shown in the following image. As you move the cursor, a preview image will appear to show the placement of the second break line.

FIGURE 5.45

7. The following image illustrates the results of creating a broken view. Notice that the dimension line appears with a break symbol to signify that the view is broken.

FIGURE 5.46

8. To edit the properties of the break view, move the cursor over the break lines, and a green circle will appear in the middle of the break. Right-click and select Edit Break from the menu. The same Break dialog box will appear. Edit the data as needed, and click the OK button to complete the edit.

9. To move the break lines, click on one of the break lines and, with the left mouse button pressed down, drag the break line to a new location, as shown in the following image. Drag the cursor away from the other break line to reduce the amount of geometry that is displayed, drag the cursor into the other break line to increase the amount geometry that is displayed. In either case the other break line will follow to maintain the gap size.

FIGURE 5.47

EXERCISE 5-3: CREATING A BREAK VIEW

In this exercise, you create a break view and a break out view.

1. Open the file *ESS_E05_03.idw*. This file has a base view and a side view that is too long to fit in the drawing.
2. You now create a break view. First click the Break command on the Place Views tab > Modify panel.
3. In the graphics window, click in the side view to break. Select the start and end points for the material to be removed, as shown in the following image.

FIGURE 5.48

4. The side view will now be broken as shown in the following image on the left.
5. To edit the properties of the break view, move the cursor over the break lines, and a green circle will appear in the middle of the break. Right-click and select Edit Break from the menu.

6. In the Break dialog box, change the Display to Max and the Gap to **0.50 inches** as shown in the following image in the middle. Click the OK button to complete the edit.
7. To move the break lines, click on one of the break lines and, with the left mouse button pressed down, drag the break line to a new location. Drag the cursor away from the other break line to reduce the amount of geometry that is displayed, drag the cursor into the other break line to increase the amount of geometry that is displayed. In either case the other break line will follow to maintain the gap size.
8. Next you will create a dimension. Press the D key on the keyboard and place the dimension on the two outside points as shown in the following image on the left. Notice the break symbol in the dimension. Dimensioning a drawing will be covered later in this chapter.
9. When done, your screen should resemble the following image on the right.

FIGURE 5.49

10. Close all open files. Do not save changes. End of exercise.

EDITING DRAWING VIEWS

After creating the drawing views, you may need to edit the properties of, delete a drawing view, or replacing the component that a drawing references. The following sections discuss these options.

Editing Drawing View Properties

After creating a drawing view, you may need to change the label, scale, style, or hatching visibility, break the alignment constraint to its base view, or control the visibility of the view projection lines for a section or auxiliary view. To edit a drawing view, follow one of these steps:

- Double-click in the bounding area of the view.
- Double-click on the icon of the view you want to edit in the browser.
- Right-click in the drawing view's bounding area or on its name, and select Edit View from the marking menu, as shown in the following image.

FIGURE 5.50

When you perform the operations listed above, the Drawing View dialog box will appear, as shown in the following image. This is the same dialog box used to create drawing views. Depending on the view that you selected, certain options may be grayed out from the dialog box. Make the necessary changes, and click the OK button to complete the edit. When you change the scale in the base view, all the dependent views will be scaled as well. You can also change the orientation of the base view; dependent views will then be updated to reflect the new orientation.

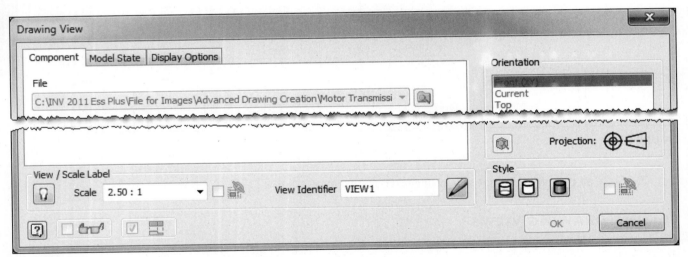

FIGURE 5.51

See the Using the Drawing View Dialog Box section in this chapter for detailed descriptions of Component and Options tab elements.

Deleting Drawing Views

To delete a drawing view, either right-click in the bounding area of the drawing view or on its name in the browser, and select Delete from the menu. You can also click in the bounding area of the drawing view, and press the DELETE key on the keyboard. A dialog box will appear, asking you to confirm the deletion of the view. If the selected view has a view that is dependent on it, you will be asked if the dependent views should also be deleted. By default, the dependent view will be deleted.

To exclude a dependent view from the delete operation, expand the dialog box, and click on the Delete cell to change the selection to No.

Break and Align Views

After creating or deleting drawing views, you may need to move a view to a projected view to a different location on the drawing but the view can only move orthographically to its parent view. To break this relationship, either click the Break Alignment command on the Place Views tab > Modify panel as shown in the following image on the left and click in the view to break alignment. The other option is to right-click in the view and click Alignment > Break from the menu as shown in the following image on the right. To align a view to another view, follow the same process but click Horizontal to align a view horizontally to a selected view, Vertical to align a view vertically to a selected view and In Position to keep a view near its current position in relation to a selected view.

FIGURE 5.52

Replace Model Reference

While annotating a drawing, you may want to change the component that the drawing reference for the geometry. Parts, assemblies, and presentation files can be replaced, but they can only be replaced with the same type of file. The file replacing the existing file should be similar in shape and size, otherwise dimensions may be orphaned. To replace a model that a drawing references, follow these steps.

1. Click the Replace Model Reference command on the Manage tab > Modify panel as shown in the following image on the left.
2. The Replace Model Reference dialog box will appear as shown in the following image on the right. All the files that are referenced in the drawing will appear in the list. In the dialog box click on the file to replace.
3. In the dialog box, click the Navigate button.
4. Browse to the new file and select it.
5. In the Open dialog box, click Open.
6. Click Yes in the warning dialog box to replace the model.
7. In the Replace Model Reference dialog box, click OK to complete the command.
8. Cleanup the drawing annotations and the views orientation as needed.

FIGURE 5.53

EXERCISE 5-4: EDITING DRAWING VIEWS

In this exercise, you will delete the base view while retaining its dependent views. The base view is not required to document the part, but its dependent views are. Next you align the section view with the right-side view to maintain the proper orthographic relationship between the views. You also modify the hatch pattern of the section to better represent the material.

1. Open the drawing file *ESS_E05_04.idw*. This drawing contains three orthographic views, an isometric view, and a section view.
2. Right-click on the base view (the middle view on the left side of the drawing), as shown in the following image, and choose Delete from the menu as shown in the following image on the left.
3. In the Delete View dialog box, notice that the projected views are highlighted on the drawing sheet. Select the More (>>) button in the dialog box.
4. Click on the word Yes in the Delete column for each dependent view to toggle each to No. This will delete the base view but leave the dependent views present on the drawing sheet as shown in the following image on the right.

FIGURE 5.54

5. In the dialog box, click OK to delete the base view and retain the two dependent views. Your drawing views should appear similar to the following image.

SECTION A-A
SCALE 1 : 1

FIGURE 5.55

With the front view deleted, the section view needs to be identified as the new base view and aligned with the right-side view.

6. Right-click in the right-side view, and click Alignment > Horizontal from the menu.
7. Select the section view as the base view.
8. Select the section view, and drag the view vertically up to the area previously occupied by the front view. Notice how the right-side view remains aligned to the section view.
9. Right-click the isometric view, and select Alignment > In Position.
10. Select the section view as the base view. Move the section view, and notice that the isometric view now moves with the section view. Your drawing should appear similar to the following image.

SECTION A-A
SCALE 1 : 1

FIGURE 5.56

You now edit the section view hatch pattern to represent the material as bronze using the ANSI 33 hatch pattern.

11. Right-click on the hatch pattern in the section view, and choose Edit from the menu.
12. The Edit Hatch Pattern dialog box is displayed. Select ANSI 33 from the Pattern list, and click OK. The hatch pattern changes, as shown in the following image.

FIGURE 5.57

Lastly you replace the part with a similar part but the holes are in different locations.

13. Click the Replace Model Reference command on the Manage tab > Modify panel.
14. In the Replace Model Reference dialog box, click on the only file in the list.
15. In the dialog box, click the Navigate button and browse to the Chapter 05 folder and select the file *ESS_E05_04_Replace.ipt* and then click Open in the dialog box.
16. In the warning dialog box, click Yes to replace the model.
17. In the Replace Model Reference dialog box, click OK to complete the command.
18. Notice from the new part there are four holes centered on the fillets but the hatch pattern and view alignment was maintained.

FIGURE 5.58

19. Close all open files. Do not save changes. End of exercise.

ANNOTATIONS

To complete an engineering drawing, you must add annotations such as centerlines, dimensions, surface texture symbols, welding symbols, geometric tolerance symbols, text, bills of materials, and balloons. How these objects appear is defined by styles. Styles will be covered as the first topic in this section.

DRAWING STANDARD AND STYLES

Each drawing is controlled by a drawing standard; the drawing standard controls overall drafting standards and which styles are available in a drawing. Only one standard can be active. Autodesk Inventor uses styles to control how objects appear. Styles control objects such as dimensions, centerlines, and text. Styles are saved within an Autodesk Inventor file or to a project library location, and they can be saved to a network location so that many users can access the same styles. This section introduces you to styles. Once a style is used in a document, it is stored in the document. Changes made to a style are only saved in that document. The updated style library can be saved to the main style library but requires permission set in the local project file and rights of the folder that the styles are saved in. To help maintain standards, it is recommended that only a limited number of people have access to write to the style library. For more information on how to save styles to a style library, consult the help system topic "Save styles to a style library".

Style Name/Value

Autodesk Inventor uses the style's name as the unique identifier of that style: only one name for the same style type can exist. For example, in a drawing, only one dimension style with a specific name "Default ANSI" can exist. An object can only be associated to one style.

Editing a Style

After dimensions have been placed, you may want to change how all of the dimension appear. Instead of changing each dimension, you can alter the dimension style or create a new style. The easiest way to edit a dimension style is to move the cursor over a dimension whose style you want to change and right-click and click Edit Dimension Style from the menu as shown in the following image.

FIGURE 5.59

Then the Style and Standard Editor dialog box will appear as shown in the following image. Make changes to the style as needed and then click the Save button on the top of the dialog box.

FIGURE 5.60

Creating a Style

Instead of editing a current dimension style, you can create a new style. To create a new style, follow these steps:

1. Click the Styles Editor command on the Manage tab > Styles and Standards panel.
2. Click and expand the style section for which you want to create a new style.
3. Right-click the style on which the new style will be based, and select New Style, as shown in the following image on the left. This image shows a new style being created from the Default (ANSI) dimension style.
4. The New Style Name dialog box will appear, as shown in the following image on the right. Enter a style name, and if you do not want the style to be used in the standard, uncheck Add to standard.
5. Make changes to the style, and save the changes.

FIGURE 5.61

6. To change existing objects to a different style, select the object or objects whose style you want to change (selected objects need to be the same object type, i.e., all dimensions), and then select a style from the Style area on the Annotate tab > Format panel as shown in the following image.

FIGURE 5.62

> Style will be covered through the exercises in this chapter.   **NOTE**

Center Marks and Centerlines

When you need to annotate the centers of holes, circular edges, or the middle (center axis) of two lines, four methods allow you to construct the needed centerlines. Use the Center Mark, Centerline, Centerline Bisector, and Centered Pattern commands under the Annotate tab, as shown in the following image. The centerlines are associated to the geometry that you select when you create them. If the geometry changes or moves, the centerlines update automatically to reflect the change. This section outlines the steps for creating the different types of centerlines.

FIGURE 5.63

Centerline

To add a centerline, follow these steps and refer to the following image:

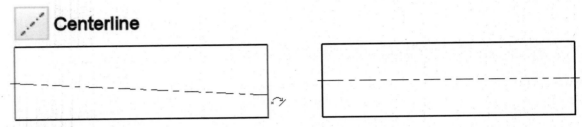

FIGURE 5.64

1. Click the Centerline command found on the Annotate tab > Symbols panel.
2. In the graphics window, select a piece of geometry for the start of the centerline.
3. Click a second piece of geometry for the ending location. You can also select multiple circles or points. If multiple points or circles are selected whose center points fall on a circle, a circle with the centerline line type (often referred to as a bolt circle) will be created.
4. Continue placing centerlines by selecting geometry.
5. Right-click, and select Create from the menu to create the centerline. The centerline will be attached to the midpoints of selected edges and center point of selected arcs and circles.
6. To complete the operation, right-click, and select Done from the menu or press the ESC key.

Centerline Bisector

To add a centerline bisector, follow the steps and refer to the following image:

FIGURE 5.65

1. Click the Centerline Bisector command found on the Annotate tab > Symbols panel.
2. In the graphics window, select two lines between which you want to place the centerline bisector. The lines do not need to be parallel to each other.
3. Continue placing centerline bisectors by selecting geometry.
4. To complete the operation, right-click, and click Done from the menu.

Center Mark

To add a center mark, follow the steps and refer to the following image:

FIGURE 5.66

5. Click the Center Mark command found on the Annotate tab > Symbols panel.
6. In the graphics window, select a center point of a circle or arc or select a circle or arc geometry in which you want to place a center mark.
7. Continue placing center marks by selecting arcs and circles.
8. To complete the operation, right-click, and click Done from the menu.

Centered Pattern

To add a centered pattern, follow the steps and refer to the following image:

FIGURE 5.67

1. Click the Centered Pattern command found on the Annotate tab > Symbols panel.
2. In the graphics window, select the circular face whose center point will be used as the center of the circle that a set of center marks will fall on, commonly referred to a bolt circle.
3. Click the first feature of the pattern.
4. Continue selecting features in a clockwise or counter clockwise direction until all of the features are added to the selection set.
5. Right-click, and select Create from the menu to create the centered pattern.
6. To complete the operation, right-click, and select Done from the menu.

Automated Centerlines from Models

Creating centerlines automatically can eliminate a considerable amount of work. You can control which features automatically get centerlines and marks and in which views these occur.

You can create automated centerlines for drawing views by right-clicking on the view boundary or hold down the CTRL key and select multiple view boundaries and right-click to display the menu and clicking Automated Centerlines from the menu as shown in the following image on the left. The Automated Centerlines dialog box will appear as shown in the following image on the right. Use this dialog box to set the type of feature(s) to which you will apply automated centerlines, such as holes, fillets, cylindrical features, etc., and to choose the projection type (plan or profile). The following sections discuss these actions in greater detail.

FIGURE 5.68

Apply To

This area controls the feature type to which you want to apply automated centerlines. Feature types include holes, fillets, cylindrical features, revolved features, circular patterns, rectangular patterns, sheet metal bends, punches, and circular sketched geometry. Click on the appropriate button to activate it, and automated centerlines will be applied to all features of that type in the drawing. You can click on multiple buttons to apply automated centerlines to multiple features. To disable centerlines in a feature, click on the feature button a second time.

Projection

Click on the projection buttons to apply automated centerlines to plan (axis normal) and/or profile (axis parallel) views.

Radius Threshold

Thresholds are minimum and maximum value settings and are provided for fillet features, arcs, and circles. Any object residing within a range should get the appropriate center mark. The values are based upon the model values, not the drawing values. This allows you to know what will or will not receive a centerline regardless of the view scale in the document. For example, if you set a minimum value of 0.50 for the fillet feature, a fillet that has a radius of 0.495 will not receive a center mark. A zero value on both threshold settings (min/max) denotes no restriction. This means that center marks will be placed on all fillets regardless of size.

Arc Angle Threshold

This option sets the minimum angle value for creating a center mark or centerline on circles, arcs, or ellipses.

> **NOTE**
> Automated centerlines can easily be created by preselecting the edges of multiple views and right-clicking the menu to display Automated Centerlines as shown in the following image.

FIGURE 5.69

EXERCISE 5-5: ADDING CENTERLINES

In this exercise, you add centerlines, notes, and move dimensions between views to a drawing.

Both model dimensions and drawing dimensions are used to document feature size.

1. Open the drawing *ESS_E05_05.dwg*. The drawing file contains four drawing views.

2. Centerlines and center marks need to be added to the drawing. Zoom in on the top view.
3. Click the Center Mark command on the Annotate tab > Symbols panel.
4. Select the left outside arc as shown in the following image on the left. Then right-click and click Done from the menu.
5. Notice that the centerline does not extend beyond the center of the circle, this is because the arc is only on the left side. Click on the centerline on the right side and drag it so it extends beyond the hidden line as shown in the following image in the middle. A center mark could have also been placed on the hole on the left side and the centerlines extended as needed.
6. Click the Centerline command on the Annotate tab > Symbols panel.
7. Select the two arcs that define the obround in the middle of the part as shown in the following image on the right. Then right-click and click Create from the menu.

 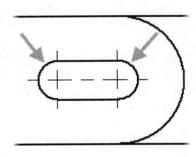

FIGURE 5.70

8. On the right-side of the top view, you create a centered pattern (bolt circle). Click the Centered Pattern command on the Annotate tab > Symbols panel.
9. First select the larger center circle labeled in the following image on the left. The center of this circle will be the center of the centered pattern. Then select the six holes that surround the circle (you can click on the circle or on the circle's center point) and then right-click and click Create from the menu. The following image on the right shows all the center lines on the top view.

FIGURE 5.71

10. Pan your screen down to show the front view.
11. Click the Centerline Bisector command on the Annotate tab > Symbols panel.
12. In the front view, select the two hidden lines that represent the drilled hole on the left side and the center hole on the right side of the part. Your screen should be similar to the following image.

FIGURE 5.72

13. In the last portion of this exercise, you create centerlines using the automated centerlines command. Close the drawing and reopen it or delete all the centerlines in the drawing.
14. Hold down the CTRL key and click in the top, front, and side view and then right-click Automated Centerlines from the menu.
15. In the Automated Centerlines dialog box, make the following selections as shown in the following image on the left:

    - In the Apply To area select: Hole Features, Fillet Features, Cylindrical Features.
    - Also in the Apply To area, select Circular Patterned Features.
    - In the Projection area, select both the Objects in View, Axis Normal and the Objects in View, Axis Parallel.
    - Click OK to create the centerlines. The following image on the right shows the three views with the automatic centerlines.

FIGURE 5.73

16. Zoom in and examine the centerlines that were created. If desired you can resize the centerlines by dragging their endpoints.
17. If the centerline and hidden lines do not display correctly, you can adjust the global line scale. To adjust the global line scale, click the Styles Editor command from the Manage tab > Styles and Standards panel.
18. In the Styles and Standard Editor on the left side, expand the Standard entry and click Default Standard (ANSI) (the bold text denotes that this is the active standard).

19. On the right side of the dialog box, enter a new value in the Global Line Scale area. To see the change in the centerlines and hidden lines in the graphics window, press ENTER on the keyboard and then click the Save button on the top of the dialog box. Try different values, when done click Done on the lower right corner of the dialog box. Styles will be covered in more detail later in this chapter.

FIGURE 5.74

20. Close all open files. Do not save changes. End of exercise.

ADDING DIMENSIONS TO A VIEW

Once you have created the drawing view(s), and centerlines you mostly likely will add dimensional annotations to the drawing. You may want to alter the model dimensions, hide certain dimensions, add drawing (general) dimensions, or move dimensions to a new location. The following sections describe these operations.

Retrieving Model Dimensions

Model dimensions may not appear automatically when you create a drawing view. You can use the Retrieve Dimensions command to select valid model dimensions for display in a drawing view. Only those dimensions that you placed in the model on a plane parallel to the view will appear.

To activate this command, use one of the following three methods:

- Click on the Retrieve command located on the Annotate tab > Dimension panel as shown in the following image on the left.
- Right-click in the bounding area of a drawing view, and select Retrieve Dimensions as shown in the following image on the right.

The appropriate dimensions for the view that were used to create the part will appear.

FIGURE 5.75

To retrieve model dimensions, follow these steps:

1. Click Retrieve on the Annotate tab > Dimension panel.
2. Select the drawing view in which to retrieve model dimensions.
3. To retrieve model dimensions based on one or more features, click the Select Features radio button. Then select the desired features.
4. If you want to retrieve model dimensions based on the entire part in a drawing view, click the Select Parts radio button. Then select the desired part or parts. You can select multiple objects by selecting them simultaneously. It is not necessary to hold the SHIFT or CTRL keys. You can also select objects with a selection window or crossing selection box.
5. Or to see all the dimensions that are available in the view, click the Select Dimensions button. If the dimensions were selected with the Feature or the model dimensions appear in the drawing view in preview mode, click the Select Dimensions button, and select the desired dimensions. The selected dimensions will be highlighted.
6. Click the Apply button to retrieve the model dimensions, and leave the dialog box active for further retrieval operations.
7. Click the OK button to retrieve the selected dimensions, and dismiss the Retrieve Dimensions dialog box.

The following image shows the effects of retrieving model dimensions using the Select Features mode. In this example, the large rectangle of the engine block was selected. This resulted in the horizontal and vertical model dimensions being retrieved.

FIGURE 5.76

The following image shows the effects of retrieving model dimensions using the Select Parts mode. In this example, the entire engine block was selected. This resulted in all model dimensions parallel to this drawing view being retrieved.

FIGURE 5.77

Changing Model Dimension Values

While working in a drawing view, you may find it necessary to change the model (parametric) dimensions of a part. You can open the part file, change a dimension's value, and save the part, and the change will then be reflected in the drawing views. You can also change a model dimension's value in the drawing view by right-clicking on the dimension and selecting Edit Model Dimension from the menu, as shown in the following image on the left.

The Edit Dimension text box will appear. Enter a new value and click the checkmark, as shown in the following image on the right, or press ENTER. The associated part will be updated and saved, and the associated drawing view(s) will be updated automatically to reflect the new value.

FIGURE 5.78

NOTE The function to edit model dimensions in a drawing is an option that can be turned on or off when you install Autodesk Inventor. However, it is considered poor practice to make major dimension changes from the drawing. It is highly recommended that you make all dimension changes through the part model.

Selecting Dimensions

As you annotate a drawing, you will need to select multiple dimensions. For example you would select multiple dimensions to arrange them, or change their dimension styles. To select multiple dimensions, you can use the same selection techniques window, crossing, etc. that you learned about in the Selecting Objects section in

Chapter 2. There are also four selection options that are available in a drawing. To use one of the selection options, click the down arrow next to the current selection command on the Quick Access toolbar as shown in the following image on the left or press down the SHIFT key and right-click and click the desired selection option from the menu as shown in the following image on the right.

FIGURE 5.79

Below are explanations for the selection options:

- **Select All Inventor Dimensions:** This option selects all dimensions that are in the current sheet.
- **Select All Model Dimensions:** This option selects all model dimensions that were retrieved in the current sheet.
- **Select All Overridden Dimensions:** This option selects all dimensions whose values have been changed or values are hidden.
- **Select All AutoCAD Dimensions:** This option is only available for *.dwg* files and selects all AutoCAD dimensions in the current sheet.

Auto-Arrange Model Dimensions

After retrieving model dimensions, the results are usually the display of dimensions that need to be rearranged in order to conform to standard dimensioning practices. You can manually reposition the dimensions or a better technique would be to have Inventor perform an automatic arrangement of the linear dimensions. To begin this process, select the dimensions using any selection technique or selection option. With all of the dimensions selected, click the Arrange command from the Annotate tab > Dimension panel as shown in the following image on the left or right-click in the graphics window and click Arrange Dimensions from the menu as shown in the following image on the right. The linear dimensions will be arranged according to the current dimension style. If needed, you will have to manually arrange dimensions like radial and diameter dimensions.

FIGURE 5.80

Moving and Centering Dimension Text

To move a dimension by either lengthening or shortening the extension lines or moving the text, position the cursor over the dimension until it becomes highlighted, as shown in the following image on the left. An icon consisting of four diagonal arrows will appear attached to your cursor. Notice the appearance of a centerline. This represents the center of the dimension text. You can drag the dimension text up, down, right, or left in order to better position it.

If you want to re-center the dimension text, slide the text toward the centerline. The dimension will snap to the centerline. In the following image on the right, the centerline changes to a dotted line to signify that the dimension text is centered. This centerline action will work on any type of linear dimension text including horizontal, vertical, and aligned. This feature is not active when repositioning the text of radius, diameter, or hole note dimensions.

FIGURE 5.81

Creating General Dimensions

After laying out the drawing views, you may find that a dimension other than an existing model dimension or an additional dimension is required to better define the part. You can go back and add a parametric dimension to the sketch if the part was underconstrained, add a driven dimension if the sketch was fully constrained, or add a general dimension to the drawing view. A general dimension is not a parametric dimension; it is associative to the geometry to which it is referenced. The general dimension reflects the size of the geometry being dimensioned. After you create a general dimension, and the value of the geometry that you dimensioned changes, the general dimension will be updated to reflect the change. You add a general dimension by using the General Dimension command found on the Annotate tab > Dimension panel as shown in the following image on the left, right-click in a blank area in the graphics screen and click the General Dimension command from the marking as shown in the following image in the middle, press the D key. You create a general dimension in the same way you would create a parametric dimension on a part's sketch. The dimensions you create in a drawing follow a similar process that you did when creating a dimension in a part's sketch, except when a drawing dimension is placed, the Edit Dimension dialog box will appear as shown in the following image on the right. In the Edit Dimension dialog box, the <<>> represents that value of the geometry being dimensioned. You can add text or symbols before or after the <<>>. You can also add a tolerance or add an inspection symbol to the dimension from the two other tabs. By default, when a drawing dimension is placed the Edit Dimension dialog box will appear. You can turn off this behavior by unchecking the Edit dimension when creating an option at the bottom of the dialog box or from the Application Options > Drawing tab. You can edit an existing drawing dimension by double-clicking on a dimension or right-clicking on a dimension and click Edit from the menu. The Edit Dimension dialog box will appear. Edit the dimension as needed.

FIGURE 5.82

Adding General Dimensions to a Drawing View

The following image on the right illustrates a simple object with various dimension types. General dimensions can take the form of linear dimensions as shown at (A) and (B), diameter dimensions as shown at (C), radius dimensions as shown at (D), and angular dimensions as shown at (E). When using the General Dimension command, Autodesk Inventor automatically chooses the dimension type depending on the object you chose. In this example, the counterbored hole is dimensioned using a special command called a hole note. This command will be described later in this chapter.

FIGURE 5.83

Adding Dimensions to Isometric Views

You can retrieve model dimensions or use of the General Dimension command to add dimensions to isometric drawing views as well as to standard orthographic views. When placing a dimension on an isometric view, the dimension text, dimension lines, extension lines, and arrows are all displayed as oblique and are aligned to the geometry being dimensioned. The following image shows a typical isometric view complete with oblique dimensions.

FIGURE 5.84

Depending on the object being dimensioned, additional controls are available to manipulate the isometric dimension being created. Once edges or points in a drawing are selected, the dimension displays based on an annotation plane, as shown in the following image on the left. If more than one annotation plane exists, you can toggle between these planes by pressing the spacebar. The results are shown in the following image on the right.

FIGURE 5.85

NOTE

You can select other annotation planes by right-clicking when placing the dimension, selecting Annotation Plane, and selecting one of the following options:

- Show All Part Work Planes
- Displays all default and user-defined work planes. These planes can then be selected as new annotation planes.
- Show All Visible Work Planes
- Displays only those work planes that are currently visible in the model. These planes can then be selected as new annotation planes.
- Use Sheet Plane
- Uses the current drawing sheet as a plane for placing the isometric dimension. Linear dimensions placed on the sheet plane are not true values.

EXERCISE 5-6: ADDING DIMENSIONS

In this exercise, you add dimensions to a drawing of a clamp and edit and create a dimensions style.

1. Open the drawing *ESS_E05_06.dwg*. The drawing file contains four drawing views with centerlines.
2. In this step, you retrieve model dimensions for use in the drawing view. Begin by zooming in on the front view.
3. Right-click in the front view, and click Retrieve Dimensions from the menu. This will launch the Retrieve Dimensions dialog box.
4. In this dialog box, click the Select Dimensions button. The model dimensions will appear on the view.
5. Drag a selection window around all dimensions. This will select the dimensions to retrieve. When finished, click the OK button. The model dimensions that are planar to the view are displayed, as shown in the following image.

FIGURE 5.86

6. Instead of manually dragging dimensions to new locations, click the Arrange command on the Annotate tab > Dimension panel.
7. Drag a selection window around all dimensions. This will select the dimensions to arrange. When finished, right-click and click Done from the menu. The dimensions are arranged according to the settings in the current dimension style.
8. There are nine dimensions that were retrieved that are not deleted. Hold down the CTRL key and select the nine dimensions that are highlighted in the following image and press the DELETE key on the keyboard or right-click and click Delete from the menu. The dimensions that were deleted can be retrieved again if needed.

FIGURE 5.87

9. Use the arrange dimension command again to arrange the dimensions.

FIGURE 5.88

10. A few of the dimensions need to be repositioned. Click and drag the dimensions to locations similar to what is shown in the following image. Note that additional dimensions could be added to the view as will be described in the next few steps.

FIGURE 5.89

> **NOTE**
>
> To reposition dimension text, make sure that no command is active.
>
> Click a dimension text object, and drag it into position. The dimension will be highlighted when it is a preset distance from the model. This distance is set in the dimension style.
>
> You can also click and drag an endpoint of an extension line; you want to do this if an extension line is going through existing lines.
>
> Radial dimensions can be repositioned by selecting the handle at the annotation end of the leader.

11. In the top view, you could retrieve dimensions. Instead you will manually add dimensions to the top view. Pan to display the top view.
12. Click the General Dimension command on the Annotate tab > Dimension panel.
13. Place the nine dimensions as shown in the following image. If needed, you can double-click on a dimension to edit a dimension's text. To change a radius dimension to a diameter dimension or vice versa, you must right-click and select the dimension type before the dimension is placed.

FIGURE 5.90

14. When the .25 TYP. horizontal dimension is created, type in **TYP.** after the <<>> as shown in the following image.

FIGURE 5.91

15. Next you change a model dimension's value via the drawing. In the front view, right-click on the 6.00 dimension and click Edit Model Dimension from the menu. Enter a value of 7 and then click on the green check mark in the Edit Dimension dialog box.

FIGURE 5.92

16. Notice how the drawing views and dimensions update to reflect the new value.
17. Next you will open the part file and change the horizontal dimension back to its original value. Move the cursor into a blank area in one of the drawing views and click Open from the menu as shown in the following image on the left.
18. Edit Sketch1 under Extrusion1 as shown in the following image on the right.

FIGURE 5.93

19. Double-click on the 7.000 horizontal dimension and enter a value **6** and then press ENTER on the keyboard.
20. Finish the sketch and then save and close the part file.
21. The drawing should be the current file; if not, make it the active file by clicking on the ESS_E05_06.dwg tab.
22. The drawing views should update to reflect the updated model.
23. Next you will change the dimension style. Move the cursor over any dimension and right-click and click Edit Dimension Style from the menu.
24. In the Style and Standard Editor, the Default (ANSI) style should be current. From the Units tab, change the Precision to 1.1 (represents one decimal place) and in the Display area, check the Leading Zeros as shown in the following image and then click Save on the top of the dialog box.

FIGURE 5.94

25. To close the dialog box, click Done on the bottom-right of the dialog box. All the dimensions should update to a single decimal place and a leading zero on the dimensions smaller than one inch in length or radius.

26. Next you create a new dimension style. Right-click on the 6.0 horizontal dimension and click New Dimension Style from the menu. The new dimension style will take on the properties of the dimension style of the selected dimension.

27. In the New Dimension Style dialog box in the upper left corner, enter a name for the new style as **3 Decimal (ANSI)**; from the Units tab change the Precision to 3.123. Click OK in the dialog box to create the dimension style.

28. The 6.0 dimension should now be 6.000 but the other dimensions are still at a single decimal place. Only 6.000 dimension is using the new 3 Decimal place dimension style.

29. Next you change a few dimensions to the new style. In the drawing, hold down the CTRL key and select a few dimensions and from the Annotate tab > Format panel select the 3 Decimal (ANSI) dimension style from the list as shown in the following image on the left.

30. Another option to change dimension to take on the properties of another dimension is to right-click on the dimension whose properties you want to copy and click Copy Properties from the menu as shown in the following image on the right. Then select the dimension(s) who you want the properties copied to. You can also right-click and click Settings from the menu and from the Copy Dimension Properties dialog box you can uncheck the properties you do not want copied.

FIGURE 5.95

31. Practice editing and creating new dimension styles.
32. In Exercise 5-8, you will continue to work on a similar drawing where you will add: hole notes, texture symbol, and a text note.
33. Close all open files. Do not save changes. End of exercise.

CREATING BASELINE DIMENSIONS

To add multiple drawing general dimensions to a drawing view in a single operation, use the Baseline Dimension command, which will place a baseline dimension about the selected geometry. The dimensions can be either horizontal or vertical. Two commands are available to assist with the creation of these dimensions: Baseline Dimension and Baseline Dimension Set.

The Baseline Dimension command allows you to create a group of horizontal or vertical dimensions from a common origin. However, the group of dimensions are not considered one group of objects; rather, each dimension that makes up a Baseline Dimension is considered a single dimension.

To create and edit baseline dimensions, follow these steps:

1. Select the Baseline Dimension command from the Annotate tab > Dimension panel as shown in the following image.

FIGURE 5.96

2. Individually select or drag a selection window around the geometry that you want to dimension. To window select, move the cursor into position where the

first point of the box will be. Press and hold down the left mouse button, and move the cursor so the preview box encompasses the geometry that you want to dimension, as shown in the following image on the left, and then release the mouse button.

3. When you are finished selecting geometry, right-click, and select Continue from the menu.
4. Move the cursor to a position where the dimensions will be placed, as shown in the following image on the right. When the dimensions are in the correct place, click the point with the left mouse button to anchor the dimensions.
5. To change the origin of the dimensions, move the cursor over the extension line that will be the origin, right-click, and click Make origin from the menu.
6. To complete the operation, right-click and select Create from the menu.
7. After creating dimensions using the Baseline Dimension command, the same editing operations can be performed on these dimensions as general dimensions. To perform these edits, move the cursor over the baseline dimension, and small green circles will appear on the dimension. Right-clicking will display the menu, as shown in the following image on the right.

FIGURE 5.97

Creating a Baseline Dimension Set

Creating a Baseline Dimension Set is similar to the Baseline Dimension as explained in the previous section. With the Baseline Dimension Set command, however, dimensions are grouped together into a set. Select the Baseline Dimension Set command from the Annotate tab > Dimension panel > click the down arrow next to the Baseline Dimension command, as shown in the following image.

FIGURE 5.98

Members can be added and deleted from the set, a new origin can be placed, and the dimensions can be automatically rearranged to reflect the change. Right-clicking on any dimension in the set will highlight all dimensions and display the menu, as shown in the following image.

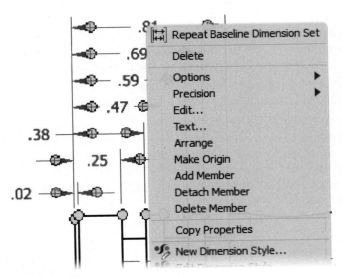

FIGURE 5.99

The main options for editing a baseline dimension set are described as follows:

Delete. This option will delete all baseline dimensions in the set. Move your cursor over any dimension in the set. When the green circles appear on each dimension, right-click, and select Delete from the menu.

Arrange. This command will rearrange the dimensions to the spacing defined in the Dimension Style.

Make Origin. This option will change the origin of the baseline dimensions. Move the cursor over an extension line of the baseline dimension set that you want to define as the new baseline, right-click, and click Make Origin from the menu.

Add Member. This option will add a drawing dimension to the baseline dimensions. Move the cursor over the baseline dimension set, when the green circles appear on the dimensions, right-click and click Add Member from the menu and then click a point that you want to add a dimension to. The dimension will be added to the set of baseline dimensions and rearranged in the proper order.

Detach Member. This option will remove a dimension from the set of baseline dimensions. Move your cursor over any dimension in the set. When the green circles appear on each dimension, right-click on the green circle on the dimension that you want to detach, right-click, and select Detach Member from the menu. The detached dimension can be moved and edited like a normal drawing dimension.

Delete Member. This option will erase a dimension from the set of baseline dimensions. Move your cursor over the dimension in the set that you want to erase, and then select Delete Member from the menu. The dimension will be deleted.

CREATING CHAIN DIMENSIONS

Another option for placing dimensions is to create chain dimensions, dimensions whose extensions lines are shared between two dimensions. Chain dimensions can be arranged horizontal or vertical. Like the baseline dimensions commands there are two Chain dimension commands, Chain and Chain Set.

The Chain command allows you to create a group of horizontal or vertical dimensions. However, the group of dimensions are not considered one group of objects; rather, each dimension that makes up the chain dimension are considered single dimensions.

To create and chain dimensions, follow these steps:

1. Select the Chain dimension command from the Annotate tab > Dimension panel as shown in the following image.

FIGURE 5.100

2. Individually select or drag a selection window around the geometry that you want to dimension. To window select, move the cursor into position where the first point of the box will be. Press and hold down the left mouse button, and move the cursor so the preview box encompasses the geometry that you want to dimension. In the following image on the left, four individual objects were selected.
3. When you are finished selecting geometry, right-click, and select Continue from the menu.
4. Move the cursor to a position where the dimensions will be placed, as shown in the following image on the right. When the dimensions are in the correct place, click the point with the left mouse button to anchor the dimensions.

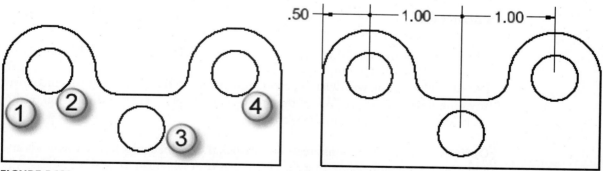

FIGURE 5.101

5. To complete the operation, right-click and select Create from the menu.
6. After creating dimensions using the Chain dimension command, the same editing operations can be performed on these dimensions as general dimensions.

Creating a Chain Dimension Set

Creating a Chain dimension Set is similar to Chain dimensions as explained in the previous section. With the Chain Set command, however, dimensions are grouped together into a set. Select the Chain Set command from the Annotate tab > Dimension panel > click the down arrow next to the Chain dimension command, as shown in the following image.

FIGURE 5.102

Members can be added and deleted from the set. Right-click on any dimension in the set will highlight all dimensions and display the menu, as shown in the following image.

FIGURE 5.103

The main options for editing a chain dimension set are described as follows:

Delete. This option will delete all chain dimensions in the set. Move your cursor over any dimension in the set. When the green circles appear on each dimension, right-click, and select Delete from the menu.

Options. From the Options area, you can lock the chain, control the arrowheads and add a leader to a dimension that the cursor was over when you right-click.

Add Member. This option will add a drawing dimension to the chain dimensions. Move the cursor over the chain dimension set, when the green circles appear on the dimensions, right-click and click Add Member from the menu and then click a point that you want to add a dimension to. The dimension will be added to the set of chain dimensions and rearranged in the proper order.

Delete Member. This option will erase a dimension from the set of chain dimensions. Move your cursor over the dimension in the set that you want to delete. When the green circles appear on each dimension, right-click, and then select Delete Member from the menu. The dimension will be deleted.

EXERCISE 5-7: CREATING BASELINE AND CHAIN DIMENSIONS

In this exercise, you will add annotations using baseline and chain dimensions.

1. Open *ESS_E05_07.dwg*.
2. You will now add a baseline dimension set. Select the Baseline Dimension Set command from the Annotate tab > Dimension panel > click the down arrow next to the Baseline Dimension command.
3. Select the three part edges as shown in the following image on the left.
4. Right-click, and click Continue from the menu.
5. Click a point above the geometry to position the dimension set as shown in the following middle image.
6. Right-click, and click Create from the menu. Notice the creation of the baseline dimension set.
7. However, all baseline dimensions need to reference the opposite edge of the object. To perform this task, move the cursor over the extension line on the right, Right-click, and click Make Origin from the menu. Notice how the dimensions are regenerated from the new origin, as shown in the following image on the right.

FIGURE 5.104

8. To add a dimension to the set, move the cursor over the baseline dimension set, when the green circles appear on the dimensions, right-click and click Add Member from the menu and then click the point as shown in the following image on the left. Right-click and click Done from the menu and the dimension will be added to the set and the dimensions will be rearranged as shown in the following image on the right.

FIGURE 5.105

9. Delete a dimension, move the cursor over the **.86** dimension, right-click and click Delete Member from the menu. The dimension will be deleted as shown in the following image on the left. Notice the dimensions were not automatically arranged.

10. Next you arrange the dimensions, move the cursor over the dimension set, when the green circles appear on the dimension right-click and click Arrange from the menu. The dimension will be rearranged as shown in the following image on the right.

FIGURE 5.106

11. Next delete the baseline dimension set. Move the cursor over the dimension set, when the green circles appear on the dimension right-click and click Delete from the menu. The dimension set will be deleted.

12. Next you create a chain dimension set. Select the Chain Set command from the Annotate tab > Dimension panel > click the down arrow next to the Chain Set dimension command.

13. Window select the part edges as shown in the following image on the left.

14. Right-click, and click Continue from the menu.

    Click a point below the geometry to position the dimension set as shown in the following image on the right.

15. Right-click, and click Create from the menu and the chain dimension set will be created.

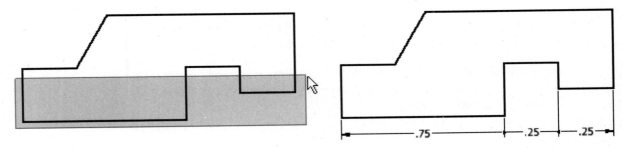

FIGURE 5.107

16. To add a dimension to the set, move the cursor over the chain dimension set, when the green circles appear on the dimensions, right-click and click Add Member from the menu and then click the point as shown in the following image on the left. Right-click and click Done from the menu. The dimension will be added to the set and the dimensions will be rearranged as shown in the following image on the right.

> **NOTE**
>
> If a dimension does not fit within its extension lines, you can right-click on the dimension and from the menu click Options > Leader and then drag the dimension and leader as needed.

FIGURE 5.108

17. Detach a dimension, move the cursor over the **.36** dimension, right-click, and click Delete Member from the menu. To center the text, right-click on the dimension set and click Arrange from the menu. The dimension will be deleted as shown in the following image. Notice the dimensions were automatically arranged.

FIGURE 5.109

18. Close all open files. Do not save changes. End of exercise.

Text and Leaders

To add text to the drawing, click either the Text or the Leader Text command on the Annotate tab > Text panel, as shown in the following image.

FIGURE 5.110

The Text command will add text while the Leader Text command will add a leader and text. Select the desired text command, and define the leader points and/or text location. Once you have chosen the location in the graphics window, the Format Text dialog box will appear. When placing text through the Format Text dialog box, select the orientation and text style as needed and type in the text, as shown in the following image. Click the OK button to place the text in the drawing. To edit the text or text leader position, move your cursor over it, click one of the green circles that appear, and drag to the desired location. To edit the text or text leader content, right-click on the text, and select Edit Leader Text or Edit Text from the menu.

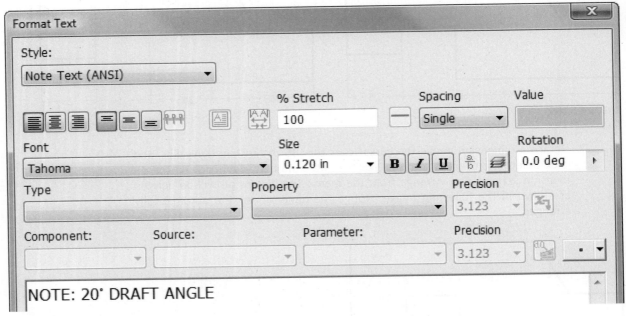

FIGURE 5.111

Additional Annotation Commands

To add more detail annotations to your drawing, you can add surface texture symbols, welding symbols, feature control frames, feature identifier symbols, datum identifier symbols, and datum targets by clicking the corresponding command on the Annotate tab > Symbols panel, as shown in the following image. The Import command is only available when you are working in an Inventor *.dwg* file.

FIGURE 5.112

Follow these steps for placing symbols:

1. Click the appropriate symbol command on the Annotate tab > Symbols panel.
2. Select a point at which the leader will start. If you don't want a leader, click a point that the symbol will be placed and proceed to step 4.
3. Continue selecting points to position the leader lines.
4. Right-click, and select Continue from the menu.
5. Fill in the information as needed in the dialog box.
6. When done, click the OK button in the dialog box.
7. To complete the operation, right-click, and select Done from the menu.
8. To edit a symbol, position the cursor over it. When the green circles appear, right-click, and click the corresponding Edit option from the menu.

Hole and Thread Notes

Another annotation you can add is a hole or thread note. Before you can place a hole or thread note in a drawing, a hole or thread feature must exist. You can also annotate extruded circles using the Hole or Thread Note command. If the hole or thread feature changes, the note will be updated automatically to reflect the change. The following image shows examples of counterbore, countersink, and threaded/tapped hole notes.

FIGURE 5.113

To create a hole or thread note, follow these steps:

1. Click the Hole/Thread Notes command on the Annotate tab > Feature Notes panel, as shown in the following image.

FIGURE 5.114

2. Select the hole or thread feature to annotate.
3. Select a second point to locate the leader and the note.
4. To complete the operation, right-click, and select Done from the menu.
5. To edit a note, position the cursor over it. When the green circles appear, right-click, and select the corresponding Edit option from the menu.

Editing Hole Notes

A hole note edit is invoked by right-clicking on an existing hole note to display the menu, as shown in the following image.

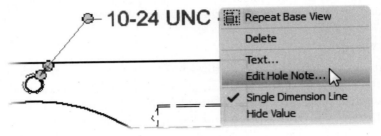

FIGURE 5.115

Selecting Edit Hole Note from this menu will display the Edit Hole Note dialog box, as shown in the following image. You can perform edits on a single hole note by unchecking the Use Default option and adding options to the hole note from the Values and Symbols area in the Edit Hole Note dialog box as shown in the following image. The following image illustrates the use of adding the <QTYNOTE> (quantity note) modifier to the hole note. This modifier will calculate the number of occurrences of the hole being noted, which is considered a highly valuable productivity command.

FIGURE 5.116

EXERCISE 5-8: ADDING ANNOTATIONS

In this exercise, you add hole notes, a surface texture symbol, and a text note.

1. Open the drawing *ESS_E05_08.dwg*. The drawing file contains four drawing views with centerlines and dimensions.
2. First you add three hole notes. Begin by zooming in on the top view.
3. Click the Hole and Thread command on the Annotate tab > Feature Notes panel.
4. Select the tapped hole on the left side and click a point to place the hole note as shown in the following image on the left.
5. Notice in the hole note there is no information about the tap drill. To add tap drill information, double-click on the hole note and in the Options area in the Edit Hole Note dialog box check the Tap Drill option as shown in the following image on the right.

FIGURE 5.117

6. Click OK to finish the edit. The following image on the left shows the updated hole note. On the right side of the top view, add two hole notes as shown in the following image on the right.

FIGURE 5.118

7. The Ø0.2 THRU hole note defines the holes that fall on the bolt circle you need to add a quantity. To add a quantity note, double-click on the Ø0.2 THRU hole note and in the Note Format area add a space after THRU and then click on Quantity Note as shown in the following image on the left.
8. Click OK to finish the edit. The following image on the right shows the updated hole note.

FIGURE 5.119

9. Next you add a surface finish note. Pan the screen so you can see the front view.
10. Click the Surface Texture Symbol command on the Annotate tab > Symbols panel.
11. Click on the top edge of the part as shown in the following image and then right-click and click Continue from the menu. Since the symbol will sit on the edge you don't need to click again to create an arrow.
12. In the Surface Texture dialog box, enter **32** in the B Production method area as shown in the following image.

FIGURE 5.120

13. Click OK to create the surface texture symbol.
14. Next you add a text note. Click the Text command on the Annotate tab > Text panel.

15. To locate the position where the text will start, click on a point below the front view and in the Format Text dialog box enter the text as shown in the following image. To add the ± symbol, select the symbol from Insert symbol area in the dialog box.

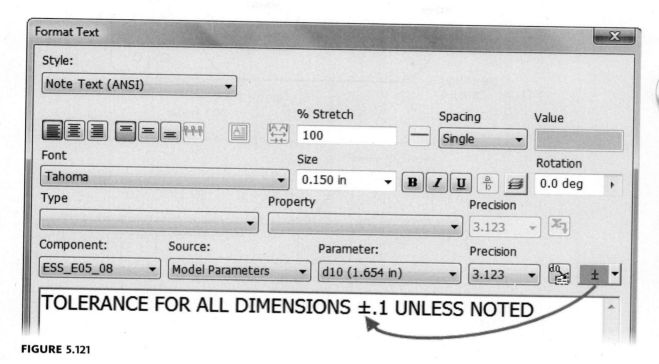

FIGURE 5.121

16. To create the text, click OK in the dialog box.
17. The text is displayed on a single line. To change the text so it appears on multiple lines, click on the text in the graphics window and then click and drag the lower-right green circle of the text's bounding area and drag it to the left as shown in the following image on the left. Release the mouse button and the text will be fitted with the new defined area as shown in the following image on the right.

FIGURE 5.122

18. Drag a selection window around all dimensions. This will select the dimensions to arrange. When finished, right-click and click Done from the menu. The dimensions are arranged according to the settings in the current dimension style.
19. When done, your screen should resemble the following image.

FIGURE 5.123

20. Close all open files. Do not save changes. End of exercise.

OPENING A MODEL FROM A DRAWING

While working inside of a complex drawing view, you may need to make changes to a part, assembly, or presentation file. Rather than close down the drawing and open the individual file, you can open a part, assembly, or presentation file directly from the drawing using a number of techniques.

One technique is to right-click in a drawing view to display the menu in the following image on the left, and then click Open to open the model file. In this example, an assembly file was opened from the drawing view.

Another technique involves expanding the drawing view in the browser to expose the assembly. Then continue expanding the browser until the file you want to open is visible in the list. Right-click the desired file in the browser and select Open from the menu to display the part model as shown in the following image on the right.

FIGURE 5.124

A third technique used to open a part file from a drawing is to first change the selection filter to Part Priority. After doing so, right-click on the specific part in the drawing view and select Open from the menu. This will open the source file.

OPENING A DRAWING FROM A MODEL

You can open a drawing from a part or assembly model by right-clicking on the part, assembly presentation file name and click Open Drawing from the menu as shown in the following image. This will open the drawing file associated with the part, assembly or presentation file. The drawing file must have the same name as the part, assembly or presentation file. Use this technique to prevent laborious searches for drawing files.

FIGURE 5.125

APPLYING YOUR SKILLS

Skill Exercise 5-1

In this exercise, you will create a drawing of a part.

1. Create a new drawing based on the *English ANSI (in).idw* template.
2. Change the sheet size to A.
3. Create drawing views from the file *ESS_Skills_5-1.ipt* from the Chapter 05 folder. Use a scale of 1:1 for all views.
4. Add centerlines, dimensions, and hole notes.

When done your drawing should resemble the following image.

FIGURE 5.126

Skill Exercise 5-2

In this exercise, you will create a drawing for a drain plate cover.

1. Begin by starting a new drawing using a English *ANSI (in).dwg* template.
2. Change the sheet size to A.
3. Create three views of the part *ESS_Skills_5-2.ipt* in the Chapter 05 folder. Use a scale of 1:1 for all views.
4. Add center marks to the views.
5. Change the Default (ANSI) dimension style:
   - Change the Extension length that the extension line goes beyond the dimension leader line to **.0625 in**.
   - Set the precision to **3.123**.
   - Set text height to **.100 in**.
6. Add dimensions and annotations. Your display should resemble the following image.

FIGURE 5.127

CHECKING YOUR SKILLS

Use these questions to test your knowledge of the material covered in this chapter.

1. True ___ False ___ A drawing can have an unlimited number of sheets.
2. Explain how to change a sheet's size.
3. True ___ False ___ There can only be one base view per sheet.
4. True ___ False ___ An inclined view is a view that is projected perpendicular to a selected edge or line in a base view.
5. True ___ False ___ An isometric view can only be projected from a base view.
6. True ___ False ___ A section view is a view created by sketching a line or multiple lines that will define the plane(s) that will be cut through a part or assembly.
7. True ___ False ___ Drawing dimensions can drive dimensional changes parametrically back to the part.
8. Explain how to shade an isometric drawing view.
9. True ___ False ___ When creating a hole note using the Hole/Thread Notes command, circles that are extruded to create a hole can be annotated.

10. True ___ False ___ An assembly model can open an individual part from its browser; however, it cannot open a drawing file from the browser.

11. True ___ False ___ When creating a section view that goes through holes at different angles, you use the projected option to prevent the holes from being distorted.

12. True ___ False ___ A hatch pattern in a section view is changed via the Document Settings.

13. True ___ False ___ Create a Segment View to create a section view that has no depth.

14. True ___ False ___ When creating a Detail View you can use a circular or rectangular fence.

15. True ___ False ___ The Styles and Standards Editor only controls dimensions styles.

16. True ___ False ___ The Centerline Bisector command can only be used to create a centerline between two parallel lines.

CHAPTER 6

Creating and Documenting Assemblies

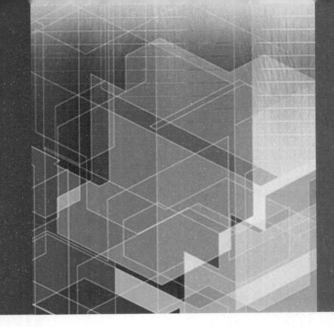

INTRODUCTION

In the first four chapters, you learned how to create a component in its own file. In this chapter, you will learn how to place individual component files into an assembly file. You will also learn to create components in the context of the assembly file. After creating components, you will learn how to constrain the components to one another using assembly constraints, edit the assembly constraints, check for interference, and create presentation files that show how the components are assembled or disassembled. Manipulating and editing a Parts List and identifying balloons.

OBJECTIVES

After completing this chapter, you will be able to perform the following:

- Understand the assembly options
- Place components into an assembly
- Create components and assemblies
- Create subassemblies
- Constrain components together using assembly constraints
- Edit assembly constraints
- Pattern components in an assembly
- Check parts in an assembly for interference
- Drive constraints
- Create a presentation file
- Create individual and automatic balloons
- Create and perform edits on a Parts List of an assembly

CREATING ASSEMBLIES

As you have already learned, component files have the *.ipt* extension, and they can only have one component each. In this chapter, you will learn how to create assembly files (*.iam* file extension). An assembly file holds the information needed to assemble the components together. All of the components in an assembly are *referenced in*, meaning that each component exists in its own component IPT file, and its definition is linked into the assembly. You can edit the components while in the assembly, or you can open the component file and edit it. When you have made changes to a component and saved the component, the changes will be reflected in the assembly after you open or update it. There are three methods for creating assemblies: bottom-up, in-place, and a combination of both. *Bottom-up* refers to an assembly in which all of the components were created in individual component files and are referenced into the assembly. The *in-place* approach refers to an assembly in which the components are created from within the context of the assembly. In other words, the user creates each component from within the top-level assembly. Each component in the assembly is saved to its own IPT file. The following sections describe the bottom-up and in-place assembly techniques.

Note that assemblies are not made up only of individual parts. Complex assemblies are typically made up of subassemblies, and therefore you can better manage the large amount of data that is created when building assemblies.

To create a new assembly, click New on the Get Started tab, and then click the Standard.iam icon, as shown in the following image. Alternately, you can click the down arrow of the Inventor icon located in the upper-left corner and click the New icon. Using either method, select Assembly from the list of templates. After issuing the new assembly operation, Autodesk Inventor's commands will change to reflect the new assembly environment. The assembly commands appear on the ribbon, and these commands are covered throughout this chapter.

> **NOTE**
> There are a number of valid approaches to creating an assembly. You will determine which method works best for the assembly that you are creating based on experience. You can create an assembly using the bottom-up technique, the in-place technique, or a combination of both. Whether you place or create the components in the assembly, all of the components will be saved to their own individual IPT files, and the assembly will be saved as an IAM file.

FIGURE 6.1

Assembly Options

Before creating an assembly, review the assembly option settings. On the Tools tab, click Application Options, and the Application Options dialog box will appear. Click on the Assembly tab, as shown in the following image. Consult the help system for more information about the Assembly Application Options. These settings are global and will affect how new components are created, referenced, analyzed, or placed in the assembly.

FIGURE 6.2

Placing Components

To place a component(s) you create the components in their individual files and then place them into an assembly. If you place an assembly file into another assembly, it will be brought in as a subassembly. Be sure to include the path(s) for the file location(s) of the placed components in the project file; otherwise, Autodesk Inventor may not be able to locate the referenced component when you reopen the assembly.

To insert a component into the current assembly, click the Place Component command on the Assemble tab, as shown in the following image on the left, press the shortcut key P, or right-click in the graphics window, and select Place Component from the marking menu.

The Place Component dialog box will appear, as shown in the following image. Select the component to place or select multiple components by holding down the CTRL or SHIFT key and select the components from the list. If the component you are placing is the first in the assembly, an instance of it will be placed into the assembly automatically. If the component is not the first, you pick a point in the assembly where you want to place the component. If you need multiple occurrences of the component in the assembly, continue selecting placement points. When done, press the ESC key, or right-click and select Done from the menu.

FIGURE 6.3

> When placing or creating components in an assembly, it is recommended to list them in the order in which they are assembled. The order is important when placing assembly constraints. To reorder a component, in the browser click and drag on a component to move it up or down.
>
> **NOTE**

CREATING PARTS IN PLACE

Most components in the assembly environment are created in relation to existing components in the assembly. When creating an in-place component, you can sketch on the face of an existing assembly component or a work plane. You can also click the graphics window background to define the current view orientation as the XY plane of the assembly in which you are presently working. If the YZ or XZ plane is the default sketch plane, you must reorient the view to see the sketch geometry. Click Application Options on the Tools tab, and then click on the Part tab to set the default sketch plane.

When you create a new component, you can select an option in the Create In-Place Component dialog box to constrain the sketch plane to the selected face or work plane automatically, as shown in the following image. After you specify the location for the sketch, the new part immediately becomes active, and the browser and ribbon switch to the part environment.

FIGURE 6.4

Notice also that the Sketch tab, as shown in the following image, is available to create sketch geometry in Sketch1 of your new part.

FIGURE 6.5

After you create the base feature of your new part, you can define additional sketches based on the active part or other parts in the assembly. When defining a new sketch, you can click a planar face of the active part or another part to define the sketch plane on that face. You can also click a planar face and drag the sketch away from the face to create the sketch plane automatically on the resulting offset work plane. When you create a sketch plane based on a face of another component, Autodesk Inventor automatically generates an adaptive work plane and places the active sketch plane on it. The adaptive work plane moves as necessary to reflect any changes in the component on which it is based. When the work plane adapts, your sketch moves with it. Features based on the sketch then adapt to match its new position.

After you finish creating a new part, you can return to assembly mode by double-clicking the assembly name in the browser. In assembly mode, assembly constraints become visible in the browser. If you selected the Constrain Sketch Plane to Selected Face option when you created your new part, a flush constraint will appear in the Assembly browser. As with all constraints, you can delete or edit this constraint at any time. No flush constraint is generated if you create a sketch by clicking in the graphics window or if the box is clear when selecting an existing part face.

The Assembly Browser

The Assembly browser displays the hierarchy of all parts and subassembly occurrences and assembly constraints in the assembly, as shown the following image. Each occurrence of a component is represented by a unique name. In the browser, you can select a component for editing, move components between assembly levels, reorder assembly components, control component status, rename components, edit assembly constraints, and manage design views and representations.

FIGURE 6.6

Occurrences

An occurrence, or instance, is a copy of an existing component and has the same name as the original component with a colon and a sequenced number. If the original component is named Bracket, for example, the next occurrence in the assembly will be Bracket:1 and a subsequent occurrence will be Bracket:2, as shown in the following image. If the original component changes, all of the component instances will reflect the change.

To create an occurrence, place the component. If the component already exists in the assembly, you can click the component's icon in the browser, and drag an additional occurrence into the assembly. You can also use the copy-and-paste method to place additional components by right-clicking on the component name in the browser, or on the component itself in the graphics window, and then right-clicking and selecting Copy from the menu. Then right-click and select Paste from the menu. You can also use the Windows shortcuts CTRL-C and CTRL-V to copy and paste the selected component. If you want an occurrence of the original component to have no relationship with its source component, make the original component active, use the Save Copy As command. This command is found by clicking the Inventor Application button located in the upper left corner of the display screen. Then click the arrow next to Save As and click Save Copy As, where you will enter a new name. The new component will have no relationship to the original, and you can place it in the assembly using the Place Component command.

FIGURE 6.7

The Assembly Capacity Meter

To provide feedback on the resources used by an assembly model, use the Assembly Capacity Meter. This meter is located in the lower-right corner of the display screen and is present only in the assembly environment. Three different types of information are displayed in the status bar of this meter. The first numbered block deals with the number of occurrences or instances in the active assembly. The second numbered block displays the total number of files open in order to display the assembly. The third block is a graphical display of the total amount of memory being used to open and work in this assembly.

FIGURE 6.8

EDITING A COMPONENT IN PLACE

To edit a component while in an assembly, activate the component. Only one component in the assembly can be active at a time. To make a component active, double-click on the component in the graphics window, or double-click on the file name or icon in the browser. Alternately, right-click on the component name in the browser or graphics window, and select Edit from the menu. Once the component is active, the other component names in the browser will appear shaded, as shown in the following image. If Component Opacity, in Application Options on the Assembly tab, is set to Active Only, the other components in the assembly will take on a faded appearance in the graphics window.

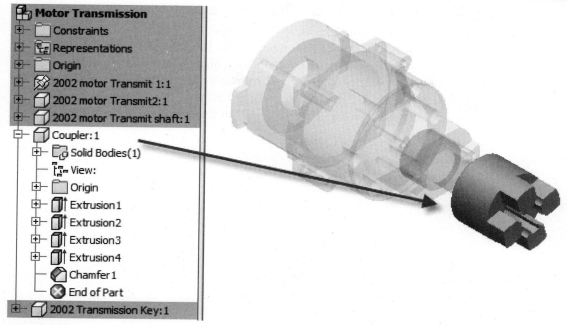

FIGURE 6.9

You can edit the component and save the changes by using the Save command. Only the active component will be saved. To make the assembly active, click the Return button on the Return panel, as shown in the following image on the left which will take you up one level, double-click on the assembly name in the browser, or right-click in the graphics window and select Finish Edit from the marking menu, as shown in the following image on the right. Under the Return button there are two other buttons, as shown in the middle of the following image; Return to Parent that will return you up one level (from the part level to the part subassembly) and Return to Top that will return you to the top level in the browser no matter how deep you are in the browser tree.

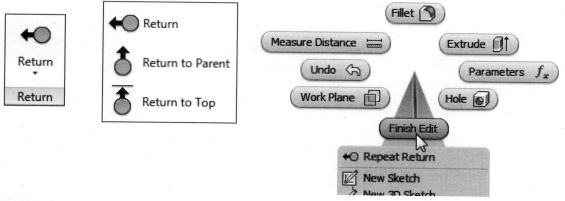

FIGURE 6.10

OPENING AND EDITING ASSEMBLY COMPONENTS

Another way to edit a component in the assembly is to open the component in another window. Click the Open command on the Inventor icon, or right-click on the component's name in the browser or on the component in the graphics window. Select Open from the menu, as shown in the following image. The component will appear in a new window. Edit the component as needed, save the changes, activate the assembly file, and the changes will appear in the assembly.

FIGURE 6.11

Grounded Components

When assembling components, you may want to make a component or multiple components grounded or stationary, meaning that they will not move. When applying assembly constraints, the unconstrained components will be moved to the grounded component(s). By default, the first component placed in an assembly is grounded. There is no limit to how many components can be grounded. It is strongly recommended that at least one component in the assembly be grounded; otherwise, the assembly can move. A grounded component is represented with a pushpin superimposed on its icon in the browser, as shown in the following image on the left. To ground or unground a component, right-click on the component's name in the browser, and select or deselect Grounded from the menu, as shown in the following image on the right.

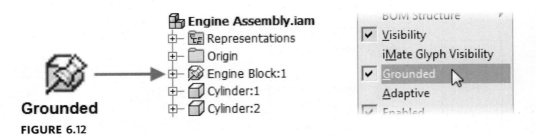

FIGURE 6.12

DEGREES OF FREEDOM (DOF)

You have now learned how to create assembly files, but the components had no relationship to one another except for the relationship that was defined when you created a component from the context of an assembly and in reference to a face on another component. For example, if you placed a bolt in a hole and the hole moved, the bolt would not move to the new hole position. Use assembly constraints to create relationships between components. With the correct constraint(s) applied, if a hole moves, the bolt will move to the new hole location.

In a previous chapter, you learned about geometric constraints. When you apply geometric constraints to sketches, they reduce the number of dimensions or constraints required to fully constrain a profile. When you apply assembly constraints, they reduce the degrees of freedom (DOF) that allow the components to move freely in space.

There are six degrees of freedom: three are translational and three are rotational. Translational means that a component can move along an axis X, Y, or Z. Rotational means that a component can rotate about an axis X, Y, or Z. As you apply assembly constraints, the number of the DOF decreases.

To see a graphical display of the DOF remaining on all of the components in an assembly, select Degrees of Freedom on the View tab, as shown in the following image.

FIGURE 6.13

An icon will appear in the center of the component that shows the DOF remaining on the component. The line and arrows represent translational freedom, and the arc and arrows represent rotational freedom. To turn off the DOF icons, again click Degrees of Freedom on the View menu.

> You can turn on the DOF symbols for single or multiple component(s) by right-clicking on the component's name in the canvas or in the browser, clicking iProperties from the menu, and clicking Degrees of Freedom on the Occurrence tab. If the symbols are turned on, following the same steps will toggle them off.

TIP

To better see the remaining degrees of freedom (DOF) you can animate them. To animate the remaining degrees of freedom (DOF) click the Degree of Freedom Analysis command from the Assemble tab > Productivity panel as shown in the following image on the left. The Degree of freedom Analysis dialog box will appear that lists the remaining degrees of freedom (DOF) as shown in the following image on the right. To animate components remaining degrees of freedom (DOF) check the Animate Freedom option and then select a component from the list in the dialog box.

FIGURE 6.14

ASSEMBLY CONSTRAINTS

When constraining components to one another, you will need to understand the terminology. The following terminology is used with assembly constraints:

Line. This can be the centerline of an arc, a circular edge, a cylindrical surface, a selected edge, a work axis, or a sketched line.

Normal. This is a vector that is perpendicular to a planar face.

Plane. This can be defined by the selection of a plane or face to include the following: two noncollinear but coplanar lines or axes, three points, or one line or axis and a point that does not lie on the line or axis. When you use edges and points to select a plane, this creates a work plane, and it is referred to as a construction plane.

Point. This can be an endpoint or midpoint of a line, the center or end of an arc or circular edge, or a vertex created by the intersection of an axis and a plane or face.

Offset. This is the distance between two selected lines, planes, or points, or any combination of the three.

Autodesk Inventor does not require components to be fully constrained. By default, the first component created or added to the assembly will be grounded and will have zero DOF. As discussed earlier, more than one component can be grounded. Other components will move in relation to the grounded component(s).

Assembly Constraint Types

Autodesk Inventor uses four types of assembly constraints (mate, angle, tangent, and insert), two types of motion constraints (rotation and rotation-translation), a transitional constraint, and a constraint set. You can access the constraints through the Constraint command found on the Assemble tab, as shown in the following image on the left, by right-clicking and selecting Constraint from the menu, or by using the hot key C.

The Place Constraint dialog box appears, as shown in the following image on the right. The dialog box is divided into four areas, which are described in the following sections. Depending upon the constraint type, the option titles may change.

FIGURE 6.15

Assembly Tab

Type. Select the type of assembly constraint to apply: mate, angle, tangent, or insert.

Selections. Click the button with the number 1, and select a component's edge, face, point, and so on, on which to base the constraint type. Then click the button with the number 2, and select a component's edge, face, point, and so on, on which to base the constraint type. By default, the second arrow will become active after you have selected the first input.

Color coding is also available to assist with the assembly process. For example, when picking a face with the number 1 button, the color blue is associated with this selection. In the same way, the color green is associated with the number 2 button selection. This schema allows you to better recognize the selections, especially if they need to be edited.

You can edit an edge, face, point, and so on, of an assembly constraint that has already been applied by clicking the number button that corresponds to the constraint and then selecting a new edge, face, point, and so on. While working on complex assemblies, you can click the box on the right side of the Selections section called Pick part first. If the box has a check, select the component before selecting a component's edge, face, point, and so on.

Offset/Angle. Enter or select a value for the offset or angle. The Offset option changes to Angle when the Angle constraint is being applied.

Solution. Select how the constraint will be applied; the normals will be pointing in the same or opposite directions.

Show Preview. Click and when constraints are applied to two components, you will see the under-constrained components previewed in their constrained positions.

If you leave the box clear, you will not see the components assembled until you click the Apply button.

Predict Offset and Orientation. Click to display the existing offset distance between two components. This allows you to accept this offset distance or enter a new offset distance in the edit box.

Motion Tab

Type. Select the type of assembly constraint to apply: rotation or rotation-translation.

Selections. Click the button with the number 1 and select a component's face or axis on which to base the constraint type. You will see a glyph in the graphics window previewing the direction of rotation motion. Click the button with the number 2, and select the component's axis or face on which to base the constraint type. A second glyph appears showing the direction of rotation.

Ratio. Enter or select a value for the ratio.

Solution. Select how the constraint type will be applied. The components will rotate in the same or opposite directions as previewed by the graphics window glyphs.

Transitional Tab

A transitional constraint will maintain contact between the two selected faces. You can use a transitional constraint between a cylindrical face and a set of tangent faces on another part.

Type. Select transitional as the type of assembly constraint to apply.

Selections. Click the First Selection button, and select the first face on the part that will be moving. Click the Second Selection button, and select a face around which the first part will be moving. If there are tangent faces, they will become chained automatically as part of the selected face.

Constraint Set Tab

This option will constrain two UCSs together.

Type. Select the type of assembly constraint to apply: UCS to UCS.

Selections. Select two UCSs on different parts.

Constraint Limits

While placing assembly constraints the geometry may not perfectly fit with the value of the constraint but may fit within a tolerance that is OK. By setting constraint limits you specify the distance or rotation that the geometry can deviate from an exact location without having the constraint fail. To set a limit for a constraint click the More button >> on the lower-right corner of the Place Constraint dialog box and the Limits section will appear as shown in the following image on the left. An assembly constraint that has limits will appear in the browser with a +/− symbol appended to its name as shown in the following image on the right.

FIGURE 6.16

Name. Enter a unique name for the constraint in the browser. If left blank a default name is used.

Use Offset or Angle as Resting Position. This option sets a value as the default position of a constraint with limits. When checked, the value entered in either the Offset or Angle field is used as the resting position.

Maximum. This option sets the maximum value that a constraint can move or rotate.

Minimum. This option sets the minimum value that a constraint can move or rotate.

Assembly Constraint Types
This section explains each of the assembly constraint types.

Mate
There are three types of mate constraints: plane, line, and point.

Mate Plane. The mate plane constraint assembles two components so that the faces on the selected planes will be planar to and opposing one another. In the following image, the mate condition is being applied; it is selected in the Solution area of the Place Constraint dialog box.

FIGURE 6.17

Mate Line. The mate line constraint assembles the edges of lines to be collinear, as shown in the following image.

FIGURE 6.18

You can also use the mate line constraint to assemble the center axis of a cylinder with a matching hole feature or axis, as shown in the following image.

FIGURE 6.19

Mate Point. The mate point constraint assembles two points, such as centers of arcs and circular edges, endpoints, and midpoints, to be coincident, as shown in the following image.

FIGURE 6.20

Mate Flush Solution

The mate flush solution constraint aligns two components so that the selected planar faces or work planes face the same direction or have their surface normals pointing in the same direction, as shown in the following image. Planar faces are the only geometry that can be selected for this constraint.

FIGURE 6.21

Angle

The angle constraint specifies the degrees between selected planes or faces or axes. The following image shows the angle constraint with two planes selected and a 30° angle applied.

Three solutions are available when placing an angle constraint: directed angle, undirected angle, and Explicit Reference Vector. The Explicit Reference Vector option requires a third selection that defines the Z axis. You can experiment with both solutions, especially when driving the angle constraint and observing the behavior of the assembly.

FIGURE 6.22

Tangent

The tangent constraint defines a tangent relationship between planes, cylinders, spheres, cones, and ruled splines. At least one of the faces selected needs to be a curve, and you can apply the tangency to the inside or outside of the curve.

The following image shows the tangent constraint applied to one outside curved face and a selected planar face, as well as the piece with the outer and inner solutions applied.

Outer Solution   Inner Solution

FIGURE 6.23

Insert

The insert constraint takes away five degrees of freedom (DOF) with one constraint, but it only works with components that have circular edges. Select the circular edges of two different components. The centerlines of the selected circles or arcs will be aligned, and a mate constraint will be applied to the planes defined by the circular edges. Circular edges define a centerline/axis and a plane. The following image shows the insert constraint with two circular edges selected and the opposed solution applied.

FIGURE 6.24

Motion Constraint Types

There are two types of motion constraints: rotation and rotation-translation, as shown in the Type section of the following image. Motion constraints allow you to simulate the motion relationships of gears, pulleys, rack and pinions, and other devices. By applying motion constraints between two or more components, you can drive one component and cause the others to move accordingly.

Both types of motion constraints are secondary constraints, which means that they define motion but do not maintain positional relationships between components. Constrain your components fully before you apply motion constraints. You can then suppress constraints that restrict the motion of the components you want to animate.

Rotation

The rotation constraint defines a component that will rotate in relation to another component by specifying a ratio for the rotation between the two components. Use this constraint for showing the relationship between gears and pulleys. Selecting the tops of the gear faces displays the rotation glyph, as shown in the following image. You may also have to change the solution type from Forward to Backward, depending on the desired results.

FIGURE 6.25

Rotation-Translation

The rotation-translation constraint defines the rotation relative to translation between components. This type of constraint is well suited for showing the relationship between rack and pinion gear assemblies. In a rack and pinion assembly, as shown in the following image, the top face of the pinion and one of the front faces of the rack are selected. You supply a distance the rack will travel based on the pitch diameter of the pinion gear, and then you can drive the constraints and test the travel distance of the mechanism.

FIGURE 6.26

Creating a Transitional Constraint

The transitional constraint specifies the intended relationship between, typically, a cylindrical part face and a contiguous set of faces on another part, such as a cam follower in a cam slot. The transitional constraint maintains contact between the faces as you slide the component along open DOF. Access this constraint type through the Transitional tab of the Place Constraint dialog box, as shown in the following image.

FIGURE 6.27

Select the moving face first on the cam as shown in the following image on the left. Next, select the transition face, as shown in the following image in the middle. The transitional face will now contact and follow the cam rotation, as shown on the right in the following image.

FIGURE 6.28

Creating a Constraint Set

If User Coordinate Systems were defined in individual part or assembly files, these UCSs can be constrained together. The buttons found under the Type and Selections areas of the dialog box allow for the selecting of individual UCSs and having the constraints be applied to the selections.

FIGURE 6.29

Selecting Geometry and the Select Other Command

After selecting the type of assembly constraint that you want to apply, the Selections button with the number 1 will become active; if it does not automatically become active, click the button. Position the cursor over the face, edge, point, and so on, to apply the first assembly constraint.

You may need to cycle through the selection set using the Select Other command, as shown in the following image, until the correct location is highlighted. Cycle by clicking on the down arrow of the command and select an option from the menu until you see the desired constraint condition, and then press the left mouse button.

FIGURE 6.30

The next step is to position the cursor over the face, edge, point, and so on, and select the second geometry input for the assembly constraint. Again, you may need to cycle through the selection set until the correct location is highlighted. If the Show Preview option is selected in the dialog box, the components will move to show how the assembly constraint will affect the components, and you will hear a snapping sound when you preview the constraint. To change either selection, click on the button with the number 1 or 2, and select the new input. Enter a value as needed for the offset or angle, and select the correct Solution option until the desired outcome appears. Click the Apply button to complete the operation. Leave the Constraint dialog box active to define subsequent constraint relationships.

Direct Select - Assembly Constraints

Instead of placing assembly constraint via a dialog box you can use the Assemble command from the Assemble tab > Position panel as shown in the following image on the left. The Assemble command can be used to place Mate, Flush, Directed Angle, Tangent, Insert, and UCS constraints by selecting geometry and the constraint types are inferred based on the selections. After starting the command a mini-toolbar displays in the graphics window as shown in the center of the following image. It allows you to choose a specific constraint option and only geometry that matches that option is selectable. For example, the Insert Opposed option will only select circular edges. To create a constraint with the Assemble command, follow these steps

1. Click the Assemble command from the Position panel.
2. If needed select a constraint option from the mini-toolbar. The Automatic option, which is also the top level command, allows you to select all types of geometry.
3. Select geometry for the first selection set (this part will move).
4. Move the cursor over your second selection. Keep moving the cursor over the geometry until the solution type is previewed then click.
5. The mini-toolbar will change. The new state of the mini-toolbar will allow you to change the solution type and enter a value for the offset or angle.
6. When the options are set either press the ENTER key, press the Apply, or the OK button on the mini-toolbar as shown in the following image on the right.

> **NOTE** The Assemble command always moves the first component to the second one. If the first component is grounded, it will move to the new location and maintain its grounded status.

FIGURE 6.31

ALT + DRAG CONSTRAINING

Another way to apply an assembly constraint is to hold down the ALT key while dragging a part edge or face to another part edge or face; no dialog box will appear. The key to dragging and applying a constraint is to select the correct area on the part. Selecting an edge will create a different type of constraint than if a face is selected. If you select a circular edge, for example, an insert constraint will be applied. To apply a constraint while dragging a part, you cannot have another command active.

To apply an assembly constraint, follow these steps:

1. While holding down the ALT key, select the face, edge, or other part on the part that will be constrained.
2. Select a planar face, linear edge, or axis to place a mate or flush constraint. Select a cylindrical face to place a tangent constraint. Select a circular edge to place an insert constraint.
3. Drag the part into position. As you drag the part over features on other parts, you will preview the constraint type. If the face you need to constrain to is behind another face, pause until the Select Other command appears. Cycle through the possible selection options, and then click to accept the selection.

To change the constraint type previewed while you drag the part, release the ALT key and press the space bar to flip to the solution direction and if needed press one of the following shortcut keys:

Key	Constraint Type
M or 1	Mate
A or 2	Angle
T or 3	Tangent
I or 4	Insert
R or 5	Rotation-Motion
S or 6	Rotation-Translation
X or 8	Transitional

NOTE A work plane can also be used as a plane with assembly constraints, a work axis can be used to define a line, and a work point can be used to define a point.

Moving and Rotating Components

Use the Move command from the Assemble tab > Position panel, as shown in the following image, to drag individual components in any direction in the viewing plane.

To perform a move operation on a component, activate the Move Component command. Click and hold the left mouse button on the component to drag it to a new location. Drop the component at its new location by releasing the button.

Moved components will follow these guidelines:

- An unconstrained component remains in the new location when moved until you constrain it to another component.
- A full or partially constrained component initially remains in the new location. When the assembly is updated, the component adjusts its location to comply with the constraints that you have already applied.

Use the Rotate command on the Assemble tab > Position panel, as shown in the following image, to rotate an individual component. This command is very useful when constraining faces that are hidden from your view.

FIGURE 6.32

Follow these steps for rotating a component in an assembly:

1. Activate the Rotate command, and select the component to rotate. Notice the appearance of the 3D rotate symbol on the selected component in the following image.

FIGURE 6.33

2. Drag your cursor until you see the desired view of the component.
   - For free rotation, click inside the Dynamic Rotate command, and drag in the desired location.
   - To rotate about the horizontal axis, click the top or bottom handle of the Dynamic Rotate command, and drag your cursor vertically.

- To rotate about the vertical axis, click the left or right handle of the Dynamic Rotate command, and drag your cursor horizontally.
- To rotate planar to the screen, hover over the rim until the symbol changes to a circle, click the rim, and drag in a circular direction.
- To change the center of rotation, click inside or outside the rim to set the new center.

Release the mouse button to drop the component into the rotated position.

NOTE If you click the Update button after moving or rotating components in an assembly, any components constrained to a grounded component will snap to their constrained positions in the new location. A fully constrained component can be moved or rotated temporarily. Once the assembly constraints are updated, the component will resolve the constraints and return to a fully constrained location/orientation.

EDITING ASSEMBLY CONSTRAINTS

After you have placed an assembly constraint, you may want to edit, suppress, or delete it to reposition the components. There are two ways to edit assembly constraints.

Both methods are executed through the browser. In the browser, activate the assembly or subassembly that contains the component that you want to edit. Expand the component name, and you will see the assembly constraints, as shown in the following image on the left. Double-click on the constraint name, and an Edit Dimension dialog box will appear, allowing you to edit the constraint offset value. You can also right-click on the assembly constraint's name in the browser and select Delete, Edit, or Suppress from the menu, as shown in the following image.

If you select Edit, the Edit Constraint dialog box will appear. If you select Suppress, the assembly constraint will not be applied. Select Drive Constraint to drive a constraint through a sequence of steps, simulating mechanical motion. Select Delete, and the assembly constraint will be deleted from the component. If you try to place or edit an assembly constraint and it cannot be applied, a warning window that explains the problem will appear. You will have to either select new options for the operation or suppress or delete another assembly constraint that conflicts with it.

NOTE When editing offset dimension values, click once with the left mouse button on the constraint with the offset value. Then change the offset value from the edit box that will appear at the bottom of the Assembly browser.

If an assembly constraint is conflicting with another, a triangular, yellow icon with an exclamation point will appear in the browser, as shown in the following image on the right. To edit a conflicting constraint in the browser, either double-click on its name or right-click on its name and select Recover from the menu, as shown in the following image on the right. The Design Doctor will appear and guide you through the steps to fix the problem.

FIGURE 6.34

Constraint Offset Value Modification

When editing a constraint offset value, use the standard value edit control. This process is similar to editing work plane offsets and sketch dimensions, and it will allow you to measure while editing constraint offset values. Either double-click or right-click on the constraint to edit in the browser, and select Modify from the menu, as shown in the following image on the left. The Edit Dimension dialog box will appear; edit the offset value.

FIGURE 6.35

> You could also click on the constraint in the browser and a dialog box appears at the bottom of the Assembly browser, enabling you to make changes to the constraint's value.

NOTE

EXERCISE 6-1: ASSEMBLING PARTS

1. Open *ESS_E06_01.iam* in the Chapter 06 folder.

FIGURE 6.36

2. Begin assembling the connector and sleeve:
   a. Right-click in the graphics window, select Home View (F6 function key).
   b. Zoom in on the small connector and sleeve.
   c. Drag the connector so that the small end is near the sleeve, as shown in the following image.

FIGURE 6.37

3. Next add a mate constraint between the centerlines of both components. Click the Constrain command from the Assemble tab > Position panel. Mate is the default constraint.

4. Move the cursor over the hole in the arm on the sleeve. Click when the centerline displays, as shown in the following image.
5. Move the cursor over the hole in the link. Click when the centerline displays, as shown in the following image.

> **NOTE** If the green dot displays, move the cursor until the centerline displays, or use Select Other to cycle through the available choices to select the axis.

6. Click Apply to accept this constraint.

FIGURE 6.38

7. Now add a mate between the faces of both parts. Click the small flat face on the ball in the link end, as shown in the following image.
8. Click the inner face of the slot on the sleeve, as shown in the following image.
9. Click Apply to accept this constraint. Click Cancel to close the Place Constraint dialog box.

FIGURE 6.39

10. Click on the small link arm, and drag it to view the effect of the two constraints.
11. Place the crank in the assembly by clicking the Place command on the Assemble tab > Component panel and opening the file *ESS_E06_01-Crank.ipt*. from the Chapter 06 folder.
12. Move the component near the spyder arm, and click to place the part, as shown in the following image. Right-click, and select Done.

FIGURE 6.40

13. To assemble the crank to the spyder, you will use the Assemble command with the Automatic option. Begin by rotating your view of the model so you can see the inner edge of a hole on the crank and the spyder as shown in the following image.

FIGURE 6.41

14. Click the Assemble command from the Assemble tab > Position panel. By default, the Automatic option is shown in the heads-up display.

FIGURE 6.42

15. Move the cursor over the hole in the arm on the crank, and click when the centerline displays.

FIGURE 6.43

16. Move the cursor over the hole on the spyder and click when the centerline is displayed as shown in the following image.

FIGURE 6.44

17. Click OK on the heads-up display or press ENTER to add the constraint.
18. Next, you will assemble the crank and link. Return to the Home View.
19. Drag the crank so you can see the affect of the constraint and to position it approximately as shown in the following image.

FIGURE 6.45

20. Drag the end of the small link arm close to the crank, as shown in the following image.
21. Zoom in to the crank and link.
22. Use the Constrain command to place a mate constraint between the centerlines of the two holes, as shown in the following image.

FIGURE 6.46

23. You will now place a claw in the assembly. Click the Place command from the Assemble tab > Component panel, and open the file *ESS_E06_01-Claw.ipt*.
24. Move the component near the end of the spyder arm, and click to place the part. Right-click, and select Done. Your display should appear similar to the following image.

FIGURE 6.47

25. Constrain the claw to the spyder. Click Constrain from the Assembly tab > Position panel.
26. Click Insert from the type area of the Place Constraint dialog box.
27. Select the hole on the end of the spyder arm. Then select the inside edge of the hole on the claw as shown in the following image.

FIGURE 6.48

28. Click OK to place the constraint and close the Place Constraint dialog box.
29. Assemble the link rod to the crank and claw. First drag the claw and link rod into the position as shown in the following image.

FIGURE 6.49

30. Next you use the Assemble command as an alternate method of placing an insert constraint between the link rod and the crank. Click the Assemble command from the Assemble tab > Position panel.
31. Select Insert – Opposed from the heads-up display drop-down list.

FIGURE 6.50

32. Select the top edge of the ball in the end of the link rod as shown in the following image on the left.
33. Select the inner edge of the hole in the crank as shown in the following image on the right.

FIGURE 6.51

34. Press ENTER or click OK to place the constraint.
35. Click the Constrain command from the Assemble tab > Position panel.
36. Add a mate constraint between the centerlines of the claw and end of the link rod.
37. Click OK to apply the constraint and close the Place Constraint dialog box.
38. Drag the claw to view the effect of the assembly constraints, as shown in the following image.

FIGURE 6.52

39. You will now add bolts and nuts to the spyder assembly. Begin by clicking the Place command on the Assemble tab > Component panel. Hold down the CTRL key and select ESS_E06_01-Bolt.ipt and ESS_E06_01-Nut.ipt. Click Open.
40. Place six nuts and bolts near their final position in the assembly, as shown in the following image. Right-click and select Done.

FIGURE 6.53

41. Insert the bolts into the assembly by clicking the Constrain command from the Assemble Tab > Position panel and in the Type area, click Insert to select an insert constraint.
42. Insert this bolt by selecting the top of the bolt's shank and then the edge of the hole, as shown in the following image.
43. Click OK to place the constraint and close the dialog box.

FIGURE 6.54

44. Use the Assemble command with either the Automatic or Insert – Opposed option to add the appropriate insert constraints to the remainder of the bolts.
45. Now add the nuts to the end of the bolts. Rotate the model so that you can see the back of the claw. Drag two nuts near their final location, as shown in the following image.
46. Use either the Constrain or Assemble command to place an insert constraint between the edge of the hole in the nut and the edge of the hole on the claw as shown in the following image on the right.

> You must select an edge on the face of the nut. Do not select the inner chamfered edge.

NOTE

FIGURE 6.55

47. Follow the same process to add the appropriate Insert constraints to the remainder of the nuts.
48. The completed assembly is displayed in the following image.

FIGURE 6.56

49. Close all open files. Do not save changes. End of exercise.

ADDITIONAL CONSTRAINT COMMANDS

Additional commands available to manipulate, navigate, and edit assembly constraints include browser views, Other Half, Constraint Offset Value Modification, Isolating Assembly Components, and Isolating Constraint Errors. The following sections describe these commands.

Browser Views

Two modes for viewing assembly information are located in the top of the browser toolbar: Assembly View and Modeling View. Assembly View, the default mode, displays assembly constraint symbols nested below both constrained components, as shown in the following image on the left. In this mode, the features used to create the part are hidden.

When Modeling View is set, assembly constraints are located in a Constraints folder at the top of the assembly tree, as shown in the following image and on the right. In this mode, the features used to create the part are displayed just as they are in the original part file.

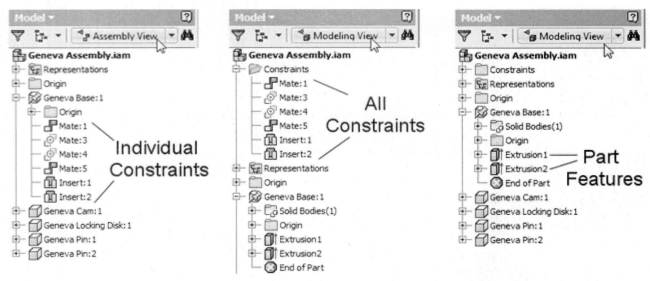

FIGURE 6.57

Other Half

You can use the Other Half command to find the matching part that participates in a constraint placed in an assembly. As you add parts over time, you may wish to highlight an assembly constraint and find the part(s) to which it is constrained. As shown in the following image on the left, a mate constraint has been highlighted. Half of this constraint has been applied to a part called Engine Block:1. To view the part sharing a common constraint, right-click on the constraint in the browser, and select Other Half from the menu, as shown in the following image on the left. The browser will expand and highlight the second half of the constraint, as shown in the following image in the middle. In this example, the other half of the mate constraint is a part called Cylinder:1.

FIGURE 6.58

Constraint Tooltip

To display all property information for a specific constraint, move your cursor over the constraint icon, and a tooltip will appear, as shown in the following image. Although the constraint name is highlighted in the image, you must hover your cursor over the constraint icon to view the tooltip information.

Mate Centerline Mate Face Angle

FIGURE 6.59

The following information is displayed in the tooltip:

- Constraint name and parameter name, applicable to offset and angle parameters
- Constrained components, that is, the two part names from the Assembly browser
- Constraint solution and type
- Constraint offset or angle value

Isolating Assembly Components

Components can be isolated as a means of viewing smaller sets of components, especially in a large assembly. In the assembly browser, click on the components to isolate, right-click to display a menu, and select Isolate. All unselected components will be set to an invisible status.

Components can be isolated based on an assembly constraint. As shown in the following image on the left, right-clicking on the Mate:7 constraint and selecting Isolate Components from the menu will display the engine block and cylinder as shown in the following image on the right.

FIGURE 6.60

To return the assembly to its previous assembled state, right-click inside the browser, and select Undo Isolate from the menu, as shown in the following image.

FIGURE 6.61

Isolating Constraint Errors

Assembly constraints can be isolated when errors in their placement occur. When editing a constraint through the Design Doctor, use the Isolate Related Failed Constraint option, as shown in the following image on the right. This action will turn off all components except those that participate in the common constraint, and it will display the Constraint dialog box, allowing you to edit the constraint with the errors.

FIGURE 6.62

ASSEMBLY BROWSER COMMANDS

Additional commands are available through the Assembly browser as a means of better controlling and managing data in an assembly file. These commands include In-Place Activation, Visibility Control, Assembly Reorder, Restructuring an Assembly, demoting and promoting assembly components, and using browser filters. A few of these commands are explained as follows.

In-Place Activation

The level of the assembly that is currently active determines whether or not you can edit components or features. You can take some actions only in the active assembly and its first-level children, while other operations are valid at all levels of the active assembly.

Double-click any subassembly or part occurrence in the browser to activate it, or right-click the occurrence in the browser and select Edit. All components not associated with the active component appear shaded in the browser, as shown in the following image on the left.

Double-clicking directly on a component in the graphics window will also activate it for editing.	NOTE

FIGURE 6.63

If you are working with a shaded display, the active component appears shaded in the graphics window, and all other components appear translucent.

You can perform the following actions on the first-level children of the active assembly:

- Delete a component
- Display the DOF of a component
- Designate a component as adaptive
- Designate a component as grounded
- Edit or delete the assembly constraints between first-level components

You can edit the features of an activated part in the assembly environment. The ribbon changes to reflect the part environment when a part is activated.

| NOTE | Double-click a parent or top-level assembly in the browser to reactivate it. |

Visibility Control

Controlling the visibility of components is critical to managing large assemblies. You may need some components only for context, or the part you need may be obscured by other components. You can change the visibility of any component in the active assembly even if the component is nested many layers deep in the assembly hierarchy. To change the visibility of a component, expand the browser until the component occurrence is visible, right-click the occurrence, and select Visibility, as shown in the following image.

| NOTE | You can also right-click on a component in the graphics window and select Visibility. |

FIGURE 6.64

ADAPTIVITY

This section will introduce you to the concept of adaptivity. To learn more about adaptivity see the Help system. Adaptivity is the Autodesk Inventor function that allows the size of a part to be determined by setting up a relationship between two parts in an assembly. Adaptivity allows under-constrained sketches—features that have undefined angles or extents, hole features, and subassemblies, which contain parts that have adaptive sketches or features—to adapt to changes. The adaptivity relationship is defined by applying assembly constraints between an adaptive sketch or feature and another part. If a sketch is fully constrained, it cannot be made adaptive. However, the extruded length or revolved angle of the part can be. A part can only be adaptive in one assembly at a time. In an assembly that has multiple placements of the same part, only one occurrence can be adaptive. The other occurrences will reflect the size of the adaptive part.

An example of adaptivity would be determining the diameter of a pin from the size of a hole. You could determine the diameter of the hole from the size of the pin, and you can turn adaptivity on and off as needed. Once a part's size is determined through adaptivity and adaptivity is no longer useful you may want to turn its adaptivity off. If you want to create adaptive features, Autodesk Inventor includes features that will speed up the process of creating them. These features are not covered in detail in this book, but can be found on the Assembly tab of the Application Options dialog box.

Under-constrained Adaptive Features

The next series of images show how parts adapt when you apply assembly constraints. In the following image, a rectangular sketch for a small plate is not dimensioned (unconstrained) along its length.

FIGURE 6.65

The extruded feature is then defined as adaptive by right-clicking on the extrusion in the browser as shown in the following image. Selecting Adaptive from the menu applies the adaptive property to the sketch and the extrusion.

FIGURE 6.66

Once you have placed the parts in an assembly, right-clicking on the component in the browser activates a menu. Selecting Adaptive from the menu adds an icon consisting of two arcs with arrows next to the component, as shown in the following image on the left. The presence of this icon in the browser means that part is now adaptive. As flush constraints are placed along the edge faces of the plates, the small plate will adapt its length to meet the length of the large plate.

The results are shown in the following image on the right.

FIGURE 6.67

ASSEMBLY SECTIONS

Assembly section views can be used to visualize portions of an assembly, especially where components are hidden or obscured within chambers. While the assembly is sectioned, part and assembly tools can be used to create or modify parts within the assembly. You can use the exposed cut edges, along with the Project Cut Edges command, to assist in sketch construction for additional features that need to be created in a part. There are four assembly section tools available on the View tab > Appearance panel as shown in the following image on the left. The other two images show a before and after view of an assembly that is using the Half Section View tool.

FIGURE 6.68

The general process for using the section tools are:

1. Select the desired assembly section tool.
2. Select any existing planar face(s) or work plane(s).
3. Specify which side of the selected section remains visible by right clicking and using the Flip Section option, shown in the following image.
4. Right-click and select Done.

You can change the position of section planes in a section view using the Virtual Movement option from the context menu, as shown in the following image. When defining the section plane, you can enter a distance in the Offset dialog box, drag the section plane in the canvas, or use the mouse wheel while your cursor is on top of the offset field in the Offset dialog box, to move the section plane. The Scroll Step Size option can be used to adjust the distance of the mouse wheel scroll.

FIGURE 6.69

EXERCISE 6-2: DESIGNING PARTS IN THE ASSEMBLY CONTEXT

In this exercise, you create a lid for a container based on the geometry of the container. This cross part sketch geometry is adaptive, and it automatically updates to reflect design changes in the container.

1. Start with an existing assembly. You will then create a new part based on cross part sketch geometry. Begin by opening the file *ESS_E06_02.iam* in the Chapter 06 folder, as shown in the following image on the left.
2. Begin the process of creating a new component in the context of an assembly by clicking the Create command from the Assemble tab > Component panel.
3. Under New File Name, enter *ESS_E06_02-Lid*.
4. Click the Browse Templates button, and in the English tab, click *Standard (in).ipt*. Click OK. There should be a checkmark beside Constrain Sketch Plane to Selected Face or Plane at the bottom of the Create In-Place Component dialog box.
5. Click the OK button to exit the dialog box, and then select the top face of the container, as shown in the following image on the right. This face becomes the new sketch plane for the new part.

FIGURE 6.70

6. You will now project all geometry contained in this face. First click the Project Geometry command from the Sketch tab > Draw panel.
7. Move the cursor onto the face of the base part until the profile of the entire face is highlighted, as shown in the following image on the left.
8. Click to project the edges. Your display should appear similar to the following image on the right.

FIGURE 6.71

9. You will now create a series of clearance holes. First zoom into a tapped hole.
10. Click the Center Point Circle command, and click the projected circle's center point.
11. Move the cursor, and click to create a circle larger than the projected circle, as shown in the following image on the left.
12. Create three additional circles centered on the remaining projected holes, as shown in the following image on the right.

FIGURE 6.72

13. All circle diameters need to be made equal. Perform this operation by clicking the Equal constraint command from the Sketch tab > Constrain panel.
14. Click one circle, and then click one of the other circles. Click the first circle again, and click a third circle. Click the first circle again, and then click the fourth circle. These steps will make all circles equal to each other in diameter.

The first entity selected is the source size that the next selection will adapt and become equal to.	**NOTE**

15. Now add a dimension to one of the circles to fully constrain the sketch. First click the Dimension command from the Sketch tab > Constrain panel.
16. Click the edge of one of the circles, and click outside the circle to place the dimension.
17. Enter **.15 in** in the dimension edit box, and press ENTER to apply the dimension. Your display should appear similar to the following image on the left.
18. Next click the Finish Sketch button to exit the Sketch environment. Click the Extrude command from the Model tab > Create panel, and select the two profiles as shown in the following image on the right.

FIGURE 6.73

19. In either the Extrude dialog box or the heads-up display, enter a distance of **.25 in**, and then click OK.
20. Click the Return command to activate the assembly. Your display should appear similar to the following image on the right.

FIGURE 6.74

21. You will now modify the container base and observe how this affects the lid. In the browser, right-click ESS_E06_02-Container:1, and then select Edit. Your display should appear similar to the following image.

FIGURE 6.75

22. In the browser, right-click Extrusion1, and then select Edit Sketch.
23. Double-click the 2 in dimension, and change the value to **2.5 in**.
24. Click the Finish Sketch command. The model should appear as shown in the following image on the left.
25. Click the Return command again to return to the assembly context. Notice that the lid adapts to the modified dimensions of the base, as shown in the following image on the right.

FIGURE 6.76

26. Close all open files. Do not save changes. End of exercise.

PATTERNING COMPONENTS

You can use the Pattern command on the Assemble tab > Component panel, as shown in the following image, when you place multiple occurrences of selected parts and subassemblies that match a feature pattern on another part (a component pattern) or that have a set of circular or rectangular part patterns in an assembly (an assembly pattern).

Three tabs are available in the Pattern Component dialog box: Associate, Rectangular, and Circular, as shown in the following image. The Associate tab is the default.

FIGURE 6.77

Associated Patterns

An associated component pattern will maintain a relationship to the feature pattern that you select. For example, a bolt is component-patterned to a part bolt-hole circular pattern that consists of four holes. If the feature pattern, the bolt-hole, changes to six holes, the bolts will move to the new locations, and two new bolts will be added for the two new holes. To create an associative component pattern, there must be a feature-based rectangular or circular pattern, and the part that will be patterned should be constrained to the parent feature, that is, the original feature that was patterned in the component. Issue the Pattern command from the Assemble tab > Component panel, and the Pattern Component dialog box will appear, as shown in the following image on the left.

By default, the Component selection option is active. Select the component, such as the cap screw, or components to pattern. Next click the Feature Pattern Select button in the dialog box and select a feature, such as a hole, that is part of the feature pattern. Do not select the parent feature. After selecting the pattern, it will highlight on the part and the pattern name will appear in the dialog box. When done, click OK to create the component pattern, as shown in the following image on the right.

FIGURE 6.78

Rectangular Patterns

A rectangular pattern operation is illustrated in the following image. When performing this operation, two edges are chosen that define the direction of the columns and rows of the pattern. Enter the number of columns and rows and then the spacing or distance between these rows and columns. The resulting pattern acts like a feature pattern. After creating it, you can edit it to change its numbers, spacing, and so on.

In the following image, because one of the part edges is inclined, work axes are turned on and used for defining the X and Y directions of the pattern.

FIGURE 6.79

Circular Patterns

Circular patterns will copy selected components in a circular direction. After selecting the component or components to pattern, select an axis direction. In the following example, this element takes the form of the centerline that acts as a pivot point for the pattern. You then enter the number of occurrences or items that will make up the pattern and the circular angle. In the following example, since the bolt component is being patterned in a full circle, enter 360° as the value for the circular angle, as shown in the following image.

FIGURE 6.80

The completed pattern then acts as a single part. If one part moves, all parts move. The patterned component will be consumed into a component pattern in the browser, and each of the part occurrences will also appear as an element that you can expand.

Pattern Editing

Edit the component pattern by selecting the pattern in the graphics window and right-clicking, or by right-clicking on the pattern's name in the browser and selecting Edit from the menu. In the browser, the component that you patterned will be consumed into a component pattern, and the part occurrences will appear as elements, which can be expanded to see the part. You can suppress an individual pattern component by right-clicking on it and selecting Suppress from the menu, as shown in the following image. The suppressed component will be grayed out and a line will be struck through it as shown in the following image on the right.

FIGURE 6.81

You can break an individual part out of the pattern by right-clicking on it and selecting Independent from the menu, as shown in the following image. Once a part is independent, it no longer has a relationship with the pattern.

FIGURE 6.82

Replacing a Component Pattern

When replacing a component in a component pattern, you replace all occurrences in the selected component pattern with a newly selected component. Expand one of the elements in the browser, and select the component to replace. With this item highlighted in the browser, right-click and select Replace from the menu, as shown in the following image on the left. The Open dialog box will appear and will enable you to select the replacement component.

The following image on the right shows the result of performing the operation to replace a pattern component. Since the use of component patterns allows for better capture of design intent, replacement of all occurrences maintains this design intent without manually replacing each component in the pattern. Overall ease of assembly use is improved, and component patterns can be completed more quickly.

FIGURE 6.83

EXERCISE 6-3: PATTERNING COMPONENTS

In this exercise, you create a fastener component pattern to match an existing hole pattern and then replace the bolt in the pattern. You complete the exercise by demoting components to create a subassembly with the cap plate and patterned fasteners.

1. Open *ESS_E06_03.iam* in the Chapter 06 folder, as shown in the following image on the left. The file contains a T pipe assembly.
2. In the browser, expand the parts *ESS_E06_03-Cap_Bolt.ipt:1* and *ESS_E06_03-Cap_Nut.ipt:1*. Notice that the parts already have insert constraints applied to them, as shown in the following image on the right.

FIGURE 6.84

3. Now create a circular pattern consisting of a collection of nuts and bolts. Click the Pattern command from the Assemble tab > Component panel.
4. Press and hold the CTRL key and select the *ESS_E06_03-Cap_Bolt.ipt:1* and *ESS_E06_03-Cap_Nut.ipt:1* parts in the browser.
5. Click the Associated Features Pattern button found under the Feature Pattern Select area of the Pattern Component dialog box. In the graphics window, point to any hole in the Cap_Plate part. When the circular pattern of holes in the cap plate is highlighted, pick the edge of one of the holes, as shown in the following image on the left.
6. You should see a preview of the component pattern, as shown in the following image on the right.

FIGURE 6.85

7. Click the OK button to create the component pattern. Expand the display of the Component Pattern 1 feature in the browser.
8. In the browser, expand Element:1 and the parts beneath it. Notice that the constraints on the original components were retained, as shown in the following image.

FIGURE 6.86

9. Now replace the bolt in the pattern with a similar but different fastener. Under Element:1 in the browser, right-click the *ESS_E06_03-Cap_Bolt,* select Component, and click Replace from the menu.

10. In the Place Component dialog box, select *ESS_E06_03-Cap_Hex_Bolt.ipt*, from the Chapter 06 folder and click Open. A warning message box will appear to notify you that some assembly constraints may be lost.

11. Click the OK button. The pattern is updated to reflect the new bolt, as shown in the following image.

NOTE The fastener in this exercise is constrained to the plate with an Insert iMate constraint. Since the replacement fastener has a similar Insert iMate, the constraint to the plate is retained.

FIGURE 6.87

12. Next, you demote a number of components; this will create a subassembly with the cap plate and component pattern.

13. Hold down CTRL while you click *ESS_E06_03-Cap_Plate:1* and Component Pattern 1 in the browser.

14. Right-click, and select Component followed by Demote from the menu. In the Create In-Place Component dialog box, enter *Cap_Kit_Assembly* in the New Component Name field. When finished, click the OK button. A warning message box appears to notify you that some assembly constraints may be lost. Click Yes.

15. In the browser, expand Cap_Kit_Assembly:1. Notice that the cap plate and component pattern are now part of the new subassembly. Notice also that the constraints on the original components were retained. Your display should appear similar to the following image.

FIGURE 6.88

16. Close all open files. Do not save changes. End of exercise.

ANALYSIS COMMANDS

Various commands are available that assist in analyzing sketch, part, and assembly models. You can calculate minimum distance between components, calculate the center of gravity of parts and assemblies, and perform interference detection.

The Minimum Distance Command

You can easily obtain the minimum distance between any components, parts, or faces in an assembly. While in the assembly, select the Measure Distance command from the Inspect tab > Measure panel as shown in the following image on the left. Then change the selection priority, depending on what you are measuring. The three available selection modes are shown in the following image on the right.

FIGURE 6.89

Identifying the Center of Gravity of an Assembly

A center of gravity placed in an assembly could be critical to the overall design process of that assembly, whether it is used in the next assembly or in the main assembly. On the View tab > Visibility panel, click the Center of Gravity command as shown in the following image on the left. The image on the right shows the center of gravity icon applied to an assembly model. This icon actually consists of a triad displaying the X, Y, and Z directions. Three selectable work planes and a selectable work point area are also available for the purpose of measuring distances and angles.

FIGURE 6.90

Interference Checking

You can check for interference in an assembly using one or two sets of objects. To check the interference among sets of stationary components, make the assembly or subassembly in question active. Then click the Analyze Interference command on the Inspect tab > Interference panel as shown in the following image on the left. The Interference Analysis dialog box will appear, as shown in the following image on the right.

FIGURE 6.91

Click on Define Set #1, and select the components that will define the first set. Click on Define Set #2, and select the components that will define the second set. A component can exist in only one set. To add or delete components from either set, select the button that defines the set that you want to edit. Click components to add to the set, or press the CTRL key while selecting components to remove from the set.

> **NOTE** Use only Define Set #1 if you want to check for interference in a single group of objects.

Once you have defined the sets, click the OK button. The order in which you selected the components has no significance. If interference is found, the Interference Detected dialog box will appear, as shown in the following image.

FIGURE 6.92

The information in the dialog box defines the X, Y, and Z coordinates of the centroid of the interfering volume. It also lists the volume of the interference and the components that interfere with one another. A temporary solid will also be created in the graphics window that represents the interference. You can copy the interference report to the clipboard or print it from the commands in the Interference Detected dialog box. When the operation is complete, click the OK button, and the interfering solid will be removed from the screen.

> **NOTE**
>
> Performing an interference detection does not fix the interfering problem; it only presents a graphical representation of the problem. After analyzing and finding an interference, edit the assembly or components to remove the interference. You can also detect interference when driving constraints.

EXERCISE 6-4: ANALYZING AN ASSEMBLY

In this exercise, you analyze a partially completed assembly for interference between parts and then check the physical properties to verify design intent.

1. Open the assembly *ESS_E06_04.iam* in the Chapter 06 folder. The linkage is displayed, as shown in the following image on the left.
2. Begin the process of checking for interferences in the assembly. First zoom into the linkages, as shown in the following image on the right.

FIGURE 6.93

3. Click Analyze Interference from the Inspect tab > Interference panel.
4. When the Analyze Interference dialog box displays, select the two components for Set #1, as shown in the following image on the left.
5. For Set #2, select the spyder, as shown in the following image on the right.

FIGURE 6.94

6. Click the OK button. Notice that the Interference Detected dialog box is displayed, and the amount of interference displays on the parts of the assembly, as shown in the following image.

FIGURE 6.95

7. Click the More (>>) button to expand the dialog box and display additional information.
8. Click the OK button to close the Interference Detected dialog box.
9. The design intent of the lifting mechanism relies on the correct selection of materials. This impacts the strength and mass of the assembly. You will now display the physical properties of the entire assembly. Return to the Home View of the assembly.
10. In the browser, right-click *ESS_E06_04.iam*, and then click iProperties, as shown following image.

FIGURE 6.96

11. When the iProperties dialog box displays, click the Physical tab.
12. Click the Update button. Notice that the physical properties of the assembly are displayed.

To get a more accurate account of the assembly properties, you can change the Requested Accuracy setting to Very High. The updating action may take some time, depending on the size of the assembly being analyzed.

13. Click the Close button to dismiss this dialog box.
14. Close all open files. Do not save changes. End of exercise.

DRIVING CONSTRAINTS

You can animate or drive mechanical motion using the Drive Constraint command. To simulate motion, an angle, mate, tangent, or insert assembly constraint must exist. You can only drive one assembly constraint at a time, but you can use equations to create relationships to drive multiple assembly constraints simultaneously. To drive a constraint, right-click on the desired constraint in the browser, and select Drive Constraint from the menu, as shown in the following image on the left.

The Drive Constraint dialog box will appear, as shown in the following image on the right. Depending on the constraint that you are driving, the units may be different. Enter a Start value; the default value is the angle or offset for the constraint. Enter a value for End and a value for Pause Delay if you want a dwell time between the steps. In the dialog box, you can also choose to create an animation (AVI or WMV) file that will record the assembly motion. You can replay the recorded file without having Autodesk Inventor installed.

FIGURE 6.97

The following sections describe the Start, End, and Pause Delay controls.

Start. Sets the start position of the offset or angle.

End. Sets the end position of the offset or angle.

Pause Delay. Sets the delay between steps. The default delay units are seconds.

Use the following Motion and video buttons to control the motion and to create a (PE) recorded video file.

▶	Forward	Drives the constraint forward, from the start position to the end position.
◀	Reverse	Drives the constraint in reverse, from the end position to the start position.
■	Pause	Temporarily stops the playback of the constraint drive sequence.
⏮	Minimum	Returns the constraint to the starting value and resets the constraint driver. This button is not available unless the constraint driver has been run.
⏪	Reverse Step	Reverses the constraint driver one step in the sequence. This button is not available unless the drive sequence has been paused.
⏩	Forward Step	Advances the constraint driver one step in the sequence. This button is not available unless the drive sequence has been paused.
⏭	Maximum	Advances the constraint sequence to the end value.
⏺	Record	Begins capturing frames at the specified rate for inclusion in an animation file named and stored to a location that you define.

To set more conditions on how the motion will behave, click the More (≪) button. This action will display the expanded Drive Constraint dialog box, as shown in the following image. The following sections describe the options in the dialog box.

FIGURE 6.98

Drive Adaptivity. Click this option to adapt the component while the constraint is driven. It only applies to assembly components for which adaptivity has been defined and enabled.

Collision Detection. Click this option to drive the constraint until a collision is detected. When interference is detected, the drive constraint will stop and the components where the collision occurs will be highlighted. It also shows the constraint value at which the collision occurred.

Increment. Determines the value that the constraint will be incremented during the animation.

Amount of Value. Increments the offset or angle value by this value for each step.

Total # of Steps. The constraint offset or angle is incremented by the same value per step based on the number of steps entered and the difference between the start and stop values.

Repetitions. Sets how the driven constraint will act when it completes a cycle and how many cycles will occur.

Start/End. Drives the constraint from the start value to the end value and resets at the start value.

Start/End/Start. Drives the constraint from the start value to the end value and then drives it in reverse from the end value to the start value.

AVI Rate. Specifies how many frames are skipped before a screen capture is taken of the motion that will become a frame in the completed AVI file.

> If you try to drive a constraint and it fails, you may need to suppress or delete another assembly constraint to allow the required component degrees of freedom. To reduce the size of a WMV or AVI movie file, reduce the screen size before creating the file and use a solid background color in the graphics window.
>
> **NOTE**

EXERCISE 6-5: DRIVING CONSTRAINTS

In this exercise, you drive an angle constraint to simulate assembly component motion. Then you drive the constraint and use collision detection to determine if components interfere.

1. Open *ESS_E06_05.iam* in the Chapter 06 folder.
2. Use the Rotate and Zoom commands to examine the assembly. When finished, right-click in the graphics window, and select Home View from the menu.
3. Now create an angle constraint between the pivot base and arm. Click the Constrain command.
4. In the Place Constraint dialog box, click the Angle button.
5. Select the Base face and then arm face, select the Directed Angle solution as shown in the following image.
6. Enter an angle of **45°**. Click the OK button.

FIGURE 6.99

7. Now drive the angle constraint you just placed. In the browser, expand *ESS_E06_05-Pivot_Base:1*.
8. Right-click the Angle constraint, and then select Drive Constraint from the menu.
9. In the Drive Constraint dialog box, enter **45.00°** as the Start value and **120.00°** as the End value.
10. Click the More (>>) button.
11. In the Increment section, enter **2.00 Deg**, as shown in the following image.
12. Click the Forward button. Notice that the arm assembly interferes with the pivot base between 45.00° and 120.00°, but Inventor continues to drive the constraint.
13. Click the Reverse button.

Chapter 6 • *Creating and Documenting Assemblies* 323

FIGURE 6.100

14. Now perform a collision detection operation. In the Drive Constraint dialog box place a check in the Collision Detection box.
15. Change the Increment value to **.1**.
16. Click the Forward button. Notice that a collision is detected at 80.7° and that the interfering parts are highlighted, as shown in the following image. Click OK to return to the Drive Constraint dialog box.

FIGURE 6.101

17. Now perform a check on the full range of motion for the arm. In the Drive Constraint dialog box, enter **80.00** for the Start value and **–80.00** for the End value.
18. Ensure that the Collision Detection option is checked.
19. Change the Increment value to **2.00**.
20. Select Start/End/Start and enter 4 in the Repetitions area of the Drive Constraint dialog box.
21. Click the Reverse button to test the full range of motion for interference. No collision is detected in this range of motion.
22. In the Drive Constraint dialog box, click the OK button. The 80.00°, which represents the maximum angle before collision, is applied to the Angle constraint, as shown in the following image.

FIGURE 6.102

23. Close all open files. Do not save changes. End of exercise.

CREATING PRESENTATION FILES

After creating an assembly, you can create drawing views based on how the parts are assembled. If you need to show the components in different positions similar to an exploded view, hide specific components, or create an animation that shows how to assemble and disassemble the components, you need to create a presentation file. Components can also be hidden and later retrieved through design view representations. This topic is covered in a later chapter. A presentation file is separate from an assembly and has a file extension of *.ipn*. The presentation file is associated with the assembly file, and changes made to the assembly file will be reflected in the presentation file. You cannot create components in a presentation file. To create a new presentation file, click New on the Get Started tab > Launch panel. When the New File dialog box appears, select the Standard.ipn icon, as shown in the following image.

FIGURE 6.103

By default, the ribbon will show the presentation commands available to you, as shown in the following image. The following sections explain these commands.

FIGURE 6.104

Create View. Click to create views of the assembly. Created views are based on the same assembly model. Once created, a view will be listed as an explosion in the browser.

Tweak Components. Click to move and/or rotate parts in the view.

Precise View Rotation. Click to rotate the view by a specified angle and direction using a dialog box.

Animate. Click to create an animation of the assembly tweaks and saved camera views; an AVI or WMV file can be output.

The basic steps for creating a presentation view are to reposition (tweak) the parts in specific directions and then create an animation, if needed. The following sections discuss these steps.

Creating Presentation Views

The first step in creating a presentation is to create a presentation view. Issue the Create View command on the Presentation tab > Create panel as shown in the following image on the left, or right-click and select Create View from the menu. The Select Assembly dialog box appears, as shown in the following image on the right. The dialog box is divided into two sections: Assembly and Explosion Method.

FIGURE 6.105

Assembly

In this section, you determine on which assembly and design view to base the presentation view.

File. If an assembly file or files are open they will appear in the drop-down list. Select an assembly file or navigate to and select an assembly file on which to base the presentation view.

Options. Click to display the File Open Options dialog box. Use this dialog box to select design view representations, positional representations, and level of detail representations.

Explosion Method

In this section, you can choose to manually or automatically explode the parts, meaning that you separate the parts a given distance.

Manual. Click so that the components will not be exploded automatically. After the presentation view is created, you can add tweaks that will move or rotate the parts.

Automatic. Click so that the parts will be exploded automatically with a given value.

Create Trails. If you clicked Automatic, click to have visible trails. Trails are lines that show how the parts are exploded.

Distance. If Automatic is selected, enter a value that the parts will be exploded.

After making your selection, click the OK button to create a presentation view. The components will appear in the graphics window and a presentation view will appear in the browser. If you clicked the Automatic option, the parts will be exploded automatically. To determine how the parts will be exploded, Autodesk Inventor analyzes the assembly constraints. If you constrained parts using the mate plane option and the arrows perpendicular to the plane (normals) for both parts were pointing outward, they will be exploded away from each other the defined distance. Think of the mating parts as two magnets that want to push themselves in opposite directions. The grounded component in the browser will be the component that stays stationary, and the other components will move away from it. If you are creating trails automatically, they will generally come from the center of the part, not necessarily from the center of the holes. After expanding all of the children in the browser, you can see numerically how the parts exploded. Once the parts are exploded, the distance is referred to as a tweak. The number by each tweak reflects the distance that the component is moved from the base component.

Tweaking Components

After creating the presentation view, you may choose to edit the tweak of a component or move or rotate the component an additional distance. To reposition the components manually is to tweak them using the Tweak command. To manually tweak a component, click the Tweak Components command on the Presentation tab > Create panel as shown in the following image on the left, press the hot key T, or right-click and select Tweak Component from the menu. The Tweak Component dialog box will appear, as shown in the following image on the right. The Tweak Component dialog box has two sections: Create Tweak and Transformations.

FIGURE 6.106

Create Tweak

In this section, you select the components to tweak, set the direction and origin of the tweak, and control trail visibility.

Direction. Determine the direction or axis of rotation for the tweak. After clicking the Direction button, select an edge, face, or feature of any component in the graphics window to set the direction triad (X, Y, and Z) for the tweak. The edge, face, or feature that you selected does not need to be on the components you are tweaking.

Components. Select the components to tweak. Click the Components button, and then click the components in the graphics window or browser to tweak. If you selected a component when you started the Create Tweak operation, it will be included in the components automatically. To remove a component from the group, press and hold the CTRL key and click the component.

Trail Origin. Set the origin for the trail. Click the Trail Origin button, and then click in the graphics window to set the origin point. If you do not specify the trail origin, it will be placed at the center of mass for the part.

Display Trails. Click if you want to see the tweak trails for the selected components. Clear the checkbox to hide the trails.

Transformations. In this section, you set the type and value of a tweak.

Linear. Click the button next to the arrow and line to move the selected components in a linear fashion.

Rotation. Click the button next to the arrow and arc to rotate the selected components.

X, Y, Z. Click the X, Y, or Z coordinate button to determine the direction for a linear tweak or the axis for a rotational tweak. Alternately, you can select the arrow on the triad that represents the X, Y, or Z direction.

Tweak Distance/Tweak Angle Field. Enter a positive or negative value for the tweak distance or rotation angle, or click a point in the graphics window and move the cursor with the mouse button depressed to set the distance.

Apply. After making all selections, click the Apply button to complete the tweak.

Edit Existing Trail. To edit an existing tweak, click the Edit Existing Trail button. Select the tweak in the graphics window, and change the desired settings.

Triad Only. Click the Triad Only option to rotate the direction triad without rotating selected components. Enter the angle of rotation, and then click the Apply button. After you rotate the triad direction, you can use it to define tweaks.

Clear. Click the Clear button to remove all of the settings and set up for another tweak.

To tweak a component, follow these steps:

1. Issue the Tweak Component command.
2. Determine the direction or axis of rotation for the tweak by clicking the Direction button and selecting an edge, face, or feature.
3. Select the components to tweak by clicking the Components button and clicking the components in the graphics window or browser that will be tweaked.
4. Select any additional trail origin points.
5. Determine whether or not you want trails to be visible.
6. Set the type of tweak to linear or rotation.
7. Click the X, Y, or Z coordinate button to determine the direction for a linear tweak or the axis for a rotational tweak.
8. Enter a value for the tweak in the text box, or select a point on the screen and drag the part into its new position.
9. Click the Apply button in the Tweak Component dialog box.

To edit an individual tweak, click on its value in the browser, as shown in the following image. Enter a new value in the text box in the lower-left corner of the browser, and press ENTER to use the new value.

FIGURE 6.107

To extend all of the tweaks the same distance, right-click on the assembly's name in the browser, and select Auto Explode from the menu, as shown in the following image on the left. Enter a value, and click the OK button, as shown in the following image on the right, and all of the tweaks will be extended the same distance. You cannot use negative values with the Auto Explode method.

FIGURE 6.108

Animation

After you have repositioned the components, you can animate the components to show how they assemble or disassemble. To create an animation, click the Animate command on the Presentation tab > Create panel as shown in the following image on the left. The Animation dialog box will appear as shown in the following image. The Animation dialog box has three sections: Parameters, Motion, and Animation Sequence, which appears under the More (<<) button.

FIGURE 6.109

Parameters

In this section, you specify the playback speed and the number of repetitions for the animation.

Interval. Set this value for the playback speed of the animation in frames. The higher the number, the greater the number of steps in the tweak and the slower the animation. A smaller number will speed up the animation.

Repetitions. Set the number of times to repeat the playback. Enter the desired number of repetitions, or use the up or down arrow to select the number.

> **NOTE** To change the number of repetitions, click the Reset button on the dialog box, and then enter a new value.

To animate components, follow these steps:

1. Issue the Animation command.
2. Set the number of repetitions.
3. Adjust the tweaks as needed.

4. Click one of the play buttons to view the animation in the graphics window.
5. To record the animation to a file, click the Record button, and then click one of the Play buttons to start recording.

Changing the Animation Sequence

Click the more button >> in the lower-right corner of the Animation dialog box to see or edit the sequence of the tweaks. In the Animation Sequence section, you can change the sequence in which the tweaks happen, select the tweak, and then select the needed operation. Expanding the Animation dialog box displays the following image. For example, in the image on the left, each component will move until its complete cycle is over before another component starts to move. In the image on the right, notice all sequence numbers have changed to a value of 1. This means when the animation begins, all components will move at the same time at the same rate of speed. Experiment with the sequencing of an exploded assembly to gain more dramatic animation results.

FIGURE 6.110

Move Up. Click to move the selected tweak up one place in the list.

Move Down. Click to move the selected tweak down one place in the list.

Group. Select a number of tweaks, and then click the Group button. When tweaks are grouped, all of the tweaks in the group will move together as you change the sequence. The group assumes the sequence order of the lowest tweak number.

Ungroup. After selecting a tweak that belongs to a group, you can click the Ungroup button, and the tweak can then be moved individually in the list. The first tweak in the group assumes a number that is one higher than the group. The remaining tweaks are numbered sequentially following the first.

 TIP You can search the Help system for *Sequence View* or *Edit Tasks and Sequences* to discover additional browser views and editing capabilities when working with Presentation files.

AUTODESK INVENTOR PUBLISHER

Although beyond the scope of this book, another product, Autodesk Inventor Publisher may be useful for you to know about as it relates to illustrating or documenting your assemblies. Publisher is an application that leverages CAD models to easily create product documentation—maintaining an associative link so that updates are easily incorporated as your designs change.

After importing your CAD data into Inventor Publisher, you will find tools allowing you to easily reposition and change the appearance of components in different process steps and views, automatically explode 3D assemblies, add annotations, and more, in order to quickly and easily communicate detailed process steps from any viewpoint.

Some of the authoring tools include:

- Component move/pull reposition tools make it easy to position individual parts and restore them to their home position.
- Storyboarding a sequence of instructions that can be animated or viewed at specified speed and transition.
- Automated and Manual tools will explode all parts or just sub-assemblies.
- Edit multiple snapshots in the storyboard at the same time, re-order, or one-step reverse the storyboard.
- Text annotations can be manually entered or inherited from the CAD model.
- Imported images and labels can be used to annotate and supplement the instructions.
- Pre-defined styles for the appearance of the assembly or individual parts can be used to obtain the required appearance.

Designs can be published for use with a wide range of applications, viewers, and media players so your documentation can be accessed by anyone. An entire document, particular storyboard or a snapshot can be selected to be published. Snapshots are published as views and storyboards are published as animations or a sequence of views. The following image shows an example of a file opened and being edited in Inventor Publisher.

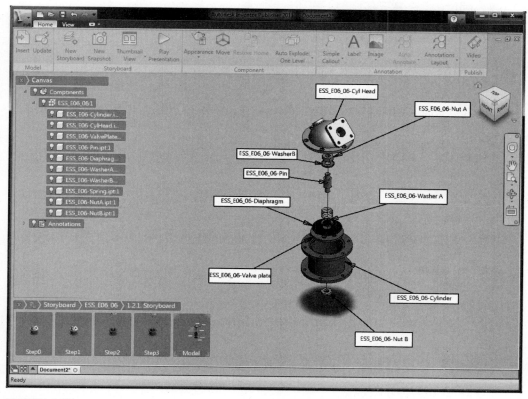

FIGURE 6.111

EXERCISE 6-6: CREATING PRESENTATION VIEWS

In this exercise, you create a presentation view of an existing assembly, and you add tweaks to the assembly components to create an exploded view.

1. Begin creating a presentation by clicking the New command.
2. Select the English tab, and then double-click *Standard (in).ipn*.
3. Click the Create View command on the Presentation tab > Create panel.
4. In the Select Assembly dialog box, verify that the Explosion Method is set to Manual. Then click the Open an Existing File button next to the File field. Open *ESS_E06_06.iam* in the Chapter 06 folder, and click the Options button in the Select Assembly dialog box.
5. In the Design View Representation field, select Internal Components, as shown in the following image and click OK.

FIGURE 6.112

6. Click OK to create the view. Your display should appear similar to the following image.

FIGURE 6.113

7. Now set the tweak direction used to create an exploded presentation. In the browser, click the plus sign in front of Explosion1 and the plus sign in front of *ESS_E06_06.iam* to expand their displays. The assembly components are displayed so you can select components in the graphics screen or in the browser.
8. Click the Tweak Components command on the Presentation tab > Create panel.
9. In the Tweak Component dialog box, verify that Direction is selected and Display Trails is checked.
10. Position your cursor over a cylindrical face until the temporary Z axis aligns with the axis of the assembly, as shown in the following image.
11. Click to accept the axis orientation.

FIGURE 6.114

12. Click on the nut at the base of the assembly to select it. In the browser, observe that *ESS_E06-NutB* is highlighted.
13. In the Tweak Components dialog box, verify that the Z button is selected, and then enter **–1.75** in the Tweak Distance field.
14. Click Apply, the green checkmark, to create the tweak. Your display should appear similar to the following image.

FIGURE 6.115

TIP	After you define the tweak axes, you can select a component or group of components and then drag them to create a tweak. This way, you can quickly arrange your components visually. You can later edit the values of the tweaks from the browser to define their final positions.

15. Next click the valve plate to select it. Hold down the CTRL key, and click the nut to deselect it. Observe that the valve plate is the only entry highlighted in the browser.
16. Enter **–0.875** in the Tweak Distance field of the dialog box.
17. Click Apply to create the tweak. Your display should appear similar to the following image.

FIGURE 6.116

> **TIP** You can also select components by clicking on them directly in the browser, which is useful when trying to select hidden components.

18. Hold down the CTRL key and click the valve plate to deselect it. The next component to tweak is the top nut. In the browser, click *ESS_E06-NutA* to select it.
19. Enter **4.5** in the Tweak Distance field, and click Apply to create the tweak. Your display should appear similar to the following image.

FIGURE 6.117

> **TIP** Examine the browser. Expand the components that have a plus sign in front of them to view the tweaks you have applied.

20. Complete this phase of the exercise by adding the following tweaks:

Part Name	Tweak
ESS_E06-WasherB.ipt:1	4
ESS_E06-Pin.ipt:1	3.125
ESS_E06-Spring.ipt:1	1.5
ESS_E06-WasherA.ipt:1	1
ESS_E06-Diaphragm.ipt:1	0.5

21. Close the Tweak Component dialog box when you are finished. Your display should appear similar to the following image.

FIGURE 6.118

22. Now adjust an existing tweak. In the browser, expand *ESS_E06-Diaphragm.ipt:1*, *ESS_E06-Spring.ipt:1*, and *ESS_E06-WasherA.ipt:1*. The hierarchy displays tweaks nested below each component definition.
23. Right-click each of the tweaks, and select Visibility to turn off the trail display for the tweak.
24. Click Tweak (0.500 in) under *ESS_E06-Diaphragm.ipt:1*.
25. Below the browser, enter a new value of **0.125** in the Offset field, and then press ENTER. Observe that the name of the tweak reflects the change and the diaphragm moves to its new position.
26. Click the Animate command from the Presentation tab > Create panel.
27. Click the Play Forward button to view the animation. When finished, click Cancel in the Animation dialog box.
28. Close all open files. Do not save changes. End of exercise.

CREATING DRAWING VIEWS FROM ASSEMBLIES AND PRESENTATION FILES

After creating an assembly or presentation, you may want to create drawings that use the data. You can create drawing views from assemblies using information from an assembly or a presentation file. You can select from them a specific design view or presentation view. To create drawing views based on assembly data, start a new drawing file or open an existing drawing file. Use the Base View command and select the IAM or IPN file from which to create the drawing. If needed, specify the design view or presentation view in the dialog box. The following image shows a presentation view being selected after a presentation file was selected.

FIGURE 6.119

CREATING BALLOONS

After you have created a drawing view of an assembly, you can add balloons to the parts and/or subassemblies. You can add balloons individually to components in a drawing, or you can use the Auto Balloon command to automatically add balloons to all components.

To add individual balloons to a drawing, follow these steps:

1. On the Annotate tab > Table panel, click the arrow below the Balloon command, as shown in the following image. Two icons will appear: the first is used

for ballooning components one at a time, and the second is used for automatic ballooning of components in a view via a single operation.

2. To balloon a single component, click the Balloon command or use the hot key B.
3. Select a component to balloon.
4. The BOM Properties dialog box will appear, as shown in the following image.

| NOTE | This dialog box will not appear if a Parts List or balloons already exist in the drawing. |

5. Select which Source and BOM Settings to use, as shown in the following image, and select OK. A preview image of the balloon will appear attached to the cursor.

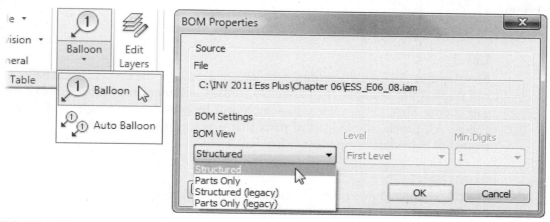

FIGURE 6.120

6. Position the cursor to place the second point for the leader, and then press the left mouse button.
7. Continue to select points to add segments to the balloon's leader, if desired.
8. When finished adding segments to the leader, right-click, and select Continue from the menu to create the balloon.
9. Select Done from the menu to cancel the operation without creating an additional balloon, or select the Back option to undo the last step that was created for the balloon.

| NOTE | When placing balloons, the direction of the balloon leader line snaps to 15° increments. The center point of a new balloon can be aligned horizontally or vertically with the center point of a balloon that already exists in a drawing. |

BOM Properties Dialog Box Options

Source
Specify the source file on which the BOM will be based.

BOM Settings and BOM View
Structured. A Structured list refers to the top-level components of an assembly in the selected view. Subassemblies and parts that belong to the subassembly will be ballooned, but parts that are in a subassembly will not be.

Parts Only View. Click to balloon only the parts of the assembly in the selected view. Subassemblies will not be ballooned, but the parts in the subassemblies will be.

> **NOTE** In a Parts Only list, components that are assemblies are not presented in the list unless they are considered inseparable or purchased.

Level
First Level or All Levels determines the level of detail that the Parts List and balloons will display. For example, if an assembly consists of a number of subassemblies and you choose First Level, each subassembly will be considered a single item. If you choose All Levels, the individual parts that make up the subassembly will also be treated as individual components when ballooned.

Min.Digits
Use this command to set the minimum amount of digits displayed for item numbering. The range is fixed from 1 to 6 digits.

Auto Ballooning
In complex assembly drawings, it will become necessary to balloon a number of components. Rather than manually add a balloon to each component, you can use the Auto Balloon command to perform this operation on several components in a single operation automatically. Clicking on the Auto Balloon button under the Annotation tab will display the Auto Balloon dialog box, as shown in the following image. The areas in this dialog box allow you to control how you place balloons.

FIGURE 6.121

Selection

This area requires you to select where to apply the balloons. With the view selected, you then add or remove components to the balloon. The Ignore Multiple Instances option appears in this area, and when the box is checked, multiple instances of the same component will not be ballooned. This action greatly reduces the number of balloon callouts in the drawing. If your application requires multiple instances to each have a balloon, remove the check from this box.

Placement

This area allows you to display the balloons along a horizontal axis, along a vertical axis, or around the view being ballooned.

BOM Settings

This area allows you to control if balloons are applied to structured, or first-level components or to parts only.

Style Overrides

The Style overrides area allows you to change the shape of the balloon and assign a user-defined balloon shape. Click Balloon Shape to enable balloon shape style overrides.

To create an auto balloon, follow these steps:

1. Select the view to which to add the balloons, as shown in the following image.

FIGURE 6.122

2. Select the components to which to add balloons. In the following image, all parts in the isometric drawing were selected using a window box. You will notice the color of all selected objects change. Balloons will be applied to these selected objects. To remove components, hold down the SHIFT or CTRL key, and click a highlighted component. This action will deselect the component. You can also select components using window or crossing window selections.

FIGURE 6.123

3. Select one of the three placement modes for the balloons. The following image shows an example of using the Horizontal Placement mode. You can further locate this balloon group by moving your cursor around the display screen. This balloon arrangement will retain its horizontal order.

FIGURE 6.124

The following image shows an example of using the Vertical Placement mode. You can further locate this balloon group by moving your cursor around the display screen. This balloon arrangement will retain its vertical order.

FIGURE 6.125

The following image shows an example of using the Around Placement mode. Move the cursor from the center of the view to readjust the balloons to different locations. In this example, the balloons are located in various positions outside of the view box. After you have placed these types of balloons, you may want to drag the balloons and arrow terminators to fine-tune individual balloon locations.

FIGURE 6.126

Editing Balloons

Edit a balloon by right-clicking on it. This action will activate a menu, as shown in the following image on the left.

Click Edit Balloon from the menu to activate the Edit Balloon dialog box, as shown in the following image on the right.

FIGURE 6.127

Use this dialog box to make changes to a balloon on an individual basis. The following features may be edited: Balloon Type and Balloon Value.

Editing the Balloon Value

The columns in the Balloon Value area of the Edit Balloon dialog box in the previous illustration specify Item and Override. When you edit the Item value, changes will be made to the balloon and will be reflected in the Parts List. In the following image, an Item was changed to 2A. Notice in the following image, on the left, that both the balloon and the Parts List have updated to the new value.

When you make a change in the Override column, the balloon will update to reflect this change, but the Parts List remains unchanged. The following image, on the right, shows this. The balloon was overridden with a value of 2A, but the Parts List has the item Rod Cap listed as 2.

FIGURE 6.128

> **NOTE** Any changes performed with the Balloon Edit dialog box will be reflected in all Parts Lists associated with that view.

Attaching a Custom Balloon

If you have predefined a custom part, you can select it from a list of custom parts for custom balloon content. To attach a custom balloon, right-click on the balloon, and select Attach Balloon From List, as shown in the following image on the left. If no custom part is available, this menu item will be grayed out.

The Attach Balloon dialog box, as shown in the following image on the right, will prompt you for your selection of custom balloons. You can select only one custom item at a time.

After selecting the custom part, click the OK button to return to the drawing. Drag the cursor around for placement of the attached balloon. Click to place the balloon in the desired location, as shown in the following image.

FIGURE 6.129

PARTS LISTS

After you have created the drawing views and placed balloons, you can also create a Parts List, as shown in the following image. As noted earlier in this section, components do not need to be ballooned before creating a Parts List.

\multicolumn{4}{c}{Parts List}	
ITEM	QTY	PART NUMBER	DESCRIPTION	MATERIAL
1	1	Arbor Press	ARBOR PRESS	Cast Steel
2	1	FACE PLATE	FACE PLATE	Steel, Mild
3	1	PINION SHAFT	PINION SHAFT	Steel, High Strength Low Alloy
4	1	LEVER ARM	LEVER ARM	Steel, Mild
5	1	THUMB SCREW	THUMB SCREW	Steel, Mild
6	1	TABLE PLATE	TABLE PLATE	Steel, High Strength Low Alloy
7	1	RAM	RAM	Steel, High Strength Low Alloy
8	2	HANDLE CAP	HANDLE CAP	Rubber
9	1	COLLAR	COLLAR	Steel, Mild
10	1	GIB PLATE	GIB PLATE	Steel, Mild
11	1	GROOVE PIN	GROOVE PIN	Steel, Mild

FIGURE 6.130

To create a Parts List, select the Annotate tab > Table panel and click the Parts List command, as shown in the following image on the left. The Parts List dialog box will appear, as shown in the following image on the right. Select a view on which to base the Parts List. After specifying your options, click the OK button, and a Parts List frame preview will appear attached to your cursor. If the components were already ballooned, the list preview will appear without displaying the Parts List dialog box. Click a point in the graphics window to place the Parts List. The information in the list is extracted from the properties of each component.

FIGURE 6.131

You can split or wrap a long Parts List into multiple columns through the Parts List dialog box. Notice the Table Wrapping area, as shown in the previous image on the right. Place a check in the Enable Automatic Wrap box to activate this feature, and then enter the number of sections to break the Parts List into. In this example, 2 is used.

The following image shows the results of performing the wrapping operation on a Parts List. Notice the two distinct sections that have been created.

Parts List				Parts List			
ITEM	QTY	PART NUMBER	DESCRIPTION	ITEM	QTY	PART NUMBER	DESCRIPTION
8	2	HANDLE CAP	HANDLE CAP	1	1	Arbor Press	ARBOR PRESS
9	1	COLLAR	COLLAR	2	1	FACE PLATE	FACE PLATE
10	1	GIB PLATE	GIB PLATE	3	1	PINION SHAFT	PINION SHAFT
11	1	GROOVE PIN	GROOVE PIN	4	1	LEVER ARM	LEVER ARM
14	4	ANSI B18.3 - 1/4 - 20 UNC - 7/8	Hex Cap Screw	5	1	THUMB SCREW	THUMB SCREW
15	4	BS 4168 - M5 x 16	Hex Cap Screws	6	1	TABLE PLATE	TABLE PLATE
16	1	ISO 4766 - M5 x 5	Hex Cap Screws	7	1	RAM	RAM

FIGURE 6.132

Editing Parts List BOM Data

Once a Parts List is created, you can open the BOM Editor directly from the drawing environment and edit the assembly BOM. All changes are saved in the assembly and the corresponding component files. Right-click on the Parts List in the graphics screen or Drawing browser, and then click Bill of Materials from the menu, as shown in the following image.

FIGURE 6.133

This action will launch the Bill of Materials editor based on a specific assembly file. The dialog box provides a convenient location to edit the iProperties and Bill of Material properties for all components in the assembly. You can even perform edits on multiple components at once. Notice in the following image that the Bill of Materials editor displays the Bill of Materials Structure. In the Bill of Materials editor, perform the desired edits. When finished, click Done to close the dialog box. All changes will be updated in the Parts List and balloons.

FIGURE 6.134

PARTS LIST COMMANDS

The information supplied earlier in this chapter covered the basics of inserting and manipulating a Parts List. This section will expand on the capabilities of the Parts List and will cover the following topics:

- A detailed account of the Parts List Operations commands available at the top of the Edit Parts List dialog box
- The function of the spreadsheet view
- Additional sections of the Edit Parts List dialog box
- Expanded capabilities of the Edit Parts List dialog box
- Nested Parts Lists

Parts List Operations (Top Icons)

Once you have placed a Parts List in the drawing, controls are available to help you modify the Parts List. Right-click on the Parts List in a drawing and select Edit Parts List from the menu, as shown in the following image, or double-click on the Parts List in the drawing.

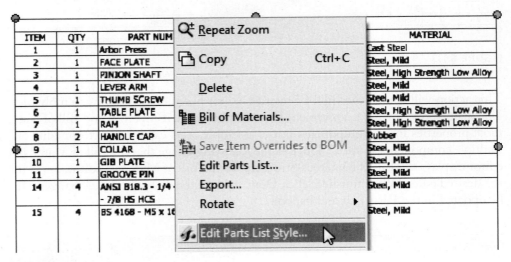

FIGURE 6.135

This action displays the Edit Parts List dialog box, as shown in the following image. The row of icons along the top of the dialog box consist of numerous features that allow you to change the Parts List.

FIGURE 6.136

The Parts List table gives detailed explanations of all Parts List icons:

Parts List Button	Function	Explanation
	Column Chooser	Opens the Parts List Column Chooser dialog box, where you can add, remove, or change the order of the columns for the selected Parts List. Data for these columns is populated from the properties of the component files.
	Group Settings	Opens the Group Settings dialog box, where you select Parts List columns to be used as a grouping key and group different components into one Parts List row. This button is active only when generating a Parts List using the Structured mode.
	Filter Settings	Opens the Filter Settings dialog box, where you can define such filter settings as Assembly View Representations, Ballooned Items Only, Item Number Range, Purchased Items, and Standard Content from which to filter the information. Once a filter is selected, you then filter out rows without changing the data in the Parts List. You can also add Parts List filters to a Parts List style.
	Sort	Opens the Sort Parts List dialog box, where you can change the sort order for items in the selected Parts List.
	Export	Opens the Export Parts List dialog box so that you can save the selected Parts List to an external file of the file type you choose.
	Table Layout	Opens the Parts List Table Layout dialog box where you can change the title text, spacing, or heading location for the selected Parts List.
	Renumber Items	Renumbers item numbers of parts in the Parts List consecutively.
	Save Item Overrides to BOM	Saves item overrides back to the assembly Bill of Materials.
	Member Selection	Used with iAssemblies. Clicking this button opens a dialog box where you can select which members of the iAssembly to include in the Parts List.

Creating Custom Parts

Custom parts are useful for displaying parts in the Parts List that are not components or graphical data, such as paint or a finishing process. To add a custom part, right-click on an existing part in the Edit Parts List dialog box, as shown in the following image, and click Insert Custom Part. When a custom part no longer needs to be documented in the Parts List, you can right-click in the row of the custom part, and click Remove Custom Part from the menu.

FIGURE 6.137

EXERCISE 6-7: CREATING ASSEMBLY DRAWINGS

In this exercise, you create drawing views of an assembly and then add balloons and a Parts List.

1. This exercise begins with formatting a drawing sheet. Open the drawing file *ESS_E06_07.idw* in the Chapter 06 folder. This drawing consists of one blank D sheet with a border and title block.
2. In the browser, right-click Sheet:1 and select Edit Sheet.
3. In the Edit Sheet dialog box, rename the sheet by typing **Assembly** in the Name field.
4. In the Size list, select C. When finished, click the OK button.
5. The next phase of this exercise involves updating the file properties in order to update the title block fields. Select the Inventor Application Menu click iProperties.
6. Click on the Summary tab and make the following changes:
   - In Title, enter **Pump Assembly**.
   - In Author, enter your name or initials.
7. Next click on the Project tab and make the following changes:
   - In Part number, enter **123-456-789**.
   - In Creation Date, click the down arrow, and then select today's date.
8. When finished, click the OK button. Zoom in to the title block and notice that the entries have updated, as shown in the following image on the left.
9. Now create drawing views. Zoom out to view the entire sheet.
10. Click the Base command on the Place Views tab > Create panel to display the Drawing View dialog box.
11. Click the Open an Existing File button next to the File list, select Assembly Files from the Files of Type drop-down list, and double-click *ESS_E06_07.iam* from the Chapter 06 folder as the view source.
12. From the Orientation list, select Top.
13. Set the Style to Hidden Line. Do not select Hidden Line Removed.
14. Set the Scale to 1:1
15. Move the view to the upper-left corner of the sheet, and then click to place the view, right-click and select Done. The drawing view should appear similar to the following image on the right.

FIGURE 6.138

16. Now begin to generate a section view from the existing top view.
17. Before you create the section view, you will need to turn off the Section property for some components so that they are not sectioned by following the next series of directions:

    - In the browser, expand the Assembly sheet.
    - In the browser, expand View1: *ESS_E06_07.iam*. Your view number may be different.
    - Expand *ESS_E06_07.iam*.
    - Select *ESS_E06-Pin.ipt*, *ESS_E06-Spring.ipt*, *ESS_E06-NutA.ipt*, and *ESS_E06-NutB.ipt*.

To select multiple items, hold down the CTRL key while clicking the items from the browser.	**TIP**

Right-click any of the highlighted parts in the browser, select Section Participation > None from the menu.

18. Now create the section view by clicking the Section command on the Place Views tab > Create panel.
19. Select the top view of the assembly, and draw a horizontal section line through the center of the top view, as shown in the image below.
20. Right-click, and select Continue.
21. Move the preview below the top view, and click to place the view.

FIGURE 6.139

22. Notice that the nuts, spring, and pin are visible in the view but not sectioned, as shown in the following image on the left. Notice also the presence of the section label and drawing view scale.
23. Now edit the section view, and turn off the label and scale. On the sheet, select the sectioned assembly view, right-click, and select Edit View from the menu to display the Drawing View dialog box.
24. Click the light bulb button (Toggle Label Visibility) to turn off the scale and label in the section view. Click the OK button. The label and scale no longer display at the bottom of the view, as shown in the following image on the right.

SECTION A-A
SCALE 1 : 1

FIGURE 6.140

25. Now hide certain components from displaying in the top view. In the browser, under the expanded component listing for *ESS_E06_07.iam*, select both *ESS_E06-CylHead:1* and *ESS_E06-Spring:1*.
26. Right-click, and clear the checkmark from Visibility. The cylinder head and the spring do not display in the top view, as shown in the following image.

FIGURE 6.141

27. Next add a text note to the top view stating that the cylinder head and spring are not visible. Click the Text command in the Annotate tab > Text panel.
28. Select a point to the left of the top assembly view.

29. In the text entry area, type **CYLINDER HEAD AND**, press ENTER, and type **SPRING REMOVED**. Click OK. Right-click and select Done. The text will be added to the top view, as shown in the following image. If necessary, select and drag the text to a better location.

FIGURE 6.142

30. Next, you create a Parts List to identify the item number, quantity, part number, and description. Begin by clicking the Parts List command on the Annotate tab > Table panel.
31. The Parts List dialog box will display. Select the sectioned assembly view.
32. Verify that the BOM View is set to Structured, and click the OK button.
33. Move the Parts List until it joins the title block and border, and then click to place it. Zoom in to display the Parts List, as shown in the following image.

Parts List			
ITEM	QTY	PART NUMBER	DESCRIPTION
1	1	Cylinder	CYLINDER
2	1	Cyl Head	CYLINDER HEAD
3	1	Valve plate	VALVE PLATE
4	1	Pin	THREADED PIN
5	1	Diaphragm	DIAPHRAGM
6	1	Washer A	RETAINING WASHER
7	1	ESS_E06_05-WasherB	RETAINING WASHER
8	1	Spring	SPRING - Ø1 X Ø22 X 25
9	1	Nut A	FLAT NUT, REG - M16 X 1.5
10	1	Nut B	FLAT NUT, THIN - M16 X 1.5

FIGURE 6.143

34. In the following steps, you will edit the Parts List to reorder the columns and adjust the column widths. Begin by selecting the Parts List in the graphics window or from the browser.
35. Right-click and select Edit Parts List from the menu to display the Parts List dialog box.

36. Click the Column Chooser button to display the Parts List Column Chooser dialog box and make the following changes:
    - In the Available Properties list, select MATERIAL.
    - Click Add.
    - In the Selected Properties list, select DESCRIPTION.
    - Click Move Up two times, and then click OK.

37. Now edit the column widths. While still in the Parts List dialog box, right-click on the DESCRIPTION header, and choose Column Width from the menu. Make the following column width changes:
    - Enter **2.25** in the Column Width dialog box.
    - Change the PART NUMBER field's width to **2**.
    - Verify the ITEM field's width is set to **1**.
    - Change the QTY field's width to **1**.
    - Change the MATERIAL field's width to **1.5**.

38. Click the OK button to close the dialog box.
39. Move the Parts List so that it is flush with the border and the title block. Your display should appear similar to the following image.

Parts List				
ITEM	DESCRIPTION	QTY	PART NUMBER	MATERIAL
1	CYLINDER	1	Cylinder	Cast Iron
2	CYLINDER HEAD	1	Cyl Head	Cast Iron
3	VALVE PLATE	1	Valve plate	Steel, Mild
4	THREADED PIN	1	Pin	Steel, Mild
5	DIAPHRAGM	1	Diaphragm	Stainless Steel
6	RETAINING WASHER	1	Washer A	Steel, Mild
7	RETAINING WASHER	1	ESS_E06_05-WasherB	Steel, Mild
8	SPRING - Ø1 X Ø22 X 25	1	Spring	Stainless Steel
9	FLAT NUT, REG - M16 X 1.5	1	Nut A	Steel, Mild
10	FLAT NUT, THIN - M16 X 1.5	1	Nut B	Steel, Mild

FIGURE 6.144

40. Now place balloons in the assembly drawing. The item number in the balloon corresponds to the item number in the Parts List.
41. Pan and zoom to display the sectioned assembly view.
42. Click the Balloon command on the Annotate tab > Table panel.
43. Select the edge of the component as the start of the leader as shown in the following image.
44. Click a point on the sheet to define the end of the first leader segment.
45. Right-click, and select Continue from the menu to place the balloon, as shown in the following image.

FIGURE 6.145

46. Select the Auto Balloon command from the Annotate tab > Table panel. Select the sectioned assembly view.
47. Window select all the components in the view. The cylinder will not be selected because it has already been ballooned.
48. Click the Select Placement button and select a location above the section view.
49. Select Horizontal for the placement and enter **0.5 in** for the Offset Spacing.
50. Click OK to place the balloons as shown in the following image.
51. Close all open files. Do not save changes. End of exercise.

FIGURE 6.146

APPLYING YOUR SKILLS

Skill Exercise 6-1

In this exercise, you create a new component for a charge pump and then assemble the pump.

1. This exercise uses the skills you have learned in previous exercises to assemble a pump and create a new part in place. Open *ESS_E06_08.iam*.

FIGURE 6.147

2. From the Chapter 06 folder place the following predefined components into the assembly:
   - 1 occurrence of *ESS_E06_08-Union.ipt*
   - 1 occurrence of *ESS_E06_08-Seal.ipt*
   - 2 occurrences of *ESS_E06_08-M8x30.ipt*

3. Next create the gland in place. Project edges from the pump body to define the flange. See the following image for the dimensions needed to complete the gland.

FIGURE 6.148

4. Use assembly constraints to build the model, as shown in the following image.

TIP
To assemble the balls, mate the center point of the ball with the centerline of the body and then place a tangent constraint between the ball's surface and the sloped surface of the seat.

To assemble the seal into the body, first apply a mate constraint between the conical surface of the seal and the conical surface of the seal's seat in the body. You must use the Select Other command to select the first conical surface. Apply a similar constraint between the gland and the seal.

FIGURE 6.149

5. The completed assembly model should appear similar to the following image.

FIGURE 6.150

6. Close all open files. Do not save changes. End of exercise.

Skill Exercise 6-2

In this exercise, you create an assembly drawing, a Parts List, and balloons for a charge pump. The charge pump assembly is shown in the following image.

FIGURE 6.151

1. Create a new drawing using the ANSI (in) DWG or IDW template with a single D sheet named Assembly.
2. Insert a top view of *ESS_E06_9.iam* with a scale of 1:1.
3. Using the top view as the base view, create a sectioned front view. Exclude the ram, valve balls, and machine screws from sectioning.
4. Use the Base command and the Change View Orientation button to create an independent top right isometric view of the pump assembly as shown in the following image.

FIGURE 6.152

5. Insert the Parts List, add a Material column and then modify its format according to the column widths shown in the following table.

Column	Width
ITEM	1
QTY	1
DESCRIPTION	1.5
PART NUMBER	2
MATERIAL	2.25

6. Modify iProperties and Parts List parameters so that the Parts List and title block match those shown in the following image. Note how the item numbers are arranged in ascending order and the header row is at the bottom of the table.

8	1	GLAND	ESS_E06_10-Gland	Brass, Soft Yellow Brass
7	2	PURCHASED	ESS_E06_09-M8x30	Default
6	1	SEAL	ESS_E06_09-Seal	Rubiconium
5	1	UNION	ESS_E06_09-Union	Rubiconium
4	1	SEAT	ESS_E06_09-Seat	Rubiconium
3	1	RAM	ESS_E06_09-Ram	Rubiconium
2	2	BALL	ESS_E06_09-Ball	Rubiconium
1	1	BODY	ESS_E06_09-Body	Rubiconium
ITEM	QTY	DESCRIPTION	PART NUMBER	MATERIAL
			PARTS LIST	

DRAWN TJ	12/24/2010			
CHECKED				
QA		TITLE		
MFG		CHARGE PUMP ASSEMBLY		
APPROVED				
		SIZE D	DWG NO	REV
		SCALE		SHEET 1 OF 1

FIGURE 6.153

7. Add balloons to identify the components as shown in the following image.

FIGURE 6.154

8. Close all open files. Do not save changes. End of exercise.

CHECKING YOUR SKILLS

Use these questions to test your knowledge of the material covered in this chapter.

1. True ___ False ___ The only way to create an assembly is by placing existing parts into it.
2. Explain how to create a component in the context of an assembly.
3. True ___ False ___ An occurrence is a new copy of an existing component.
4. True ___ False ___ Only one component can be grounded in an assembly.
5. True ___ False ___ Autodesk Inventor does not require components in an assembly to be fully constrained.
6. True ___ False ___ A sketch must be fully constrained to adapt.
7. What is the purpose of creating a presentation file?
8. True ___ False ___ Balloons can only be placed in a drawing after placing a Parts List.
9. True ___ False ___ When creating drawing views from an assembly, you can create views from multiple presentation views or design views.
10. True ___ False ___ A presentation file can be saved as a Flash animation.
11. True ___ False ___ An Associated Component Pattern will maintain a relationship to a feature pattern.
12. True ___ False ___ In a Parts List, you can display custom parts that are not components or graphical data, e.g. paint.

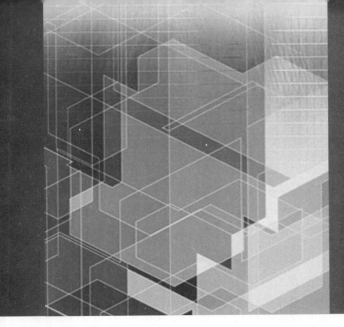

CHAPTER 7

Advanced Modeling Techniques

INTRODUCTION

In this chapter, you will learn how to use advanced modeling techniques. Using advanced modeling techniques, you can create transitions between parts that would otherwise be difficult to create. You can edit advanced features, such as other sketched and placed features, but typically their creation requires more than one unconsumed sketch. This chapter also introduces you to techniques that will help you be more productive in your modeling.

OBJECTIVES

After completing this chapter, you will be able to perform the following:

- Section a part
- Create a design view representation in a part file
- Emboss text and profiles
- Create relationships between dimensions
- Create sweep features
- Create 3D sketches
- Create loft features
- Mirror model features
- Suppress features of a part
- Reorder part features
- Understand the Content Center and Design Accelerator functionality
- Simulate a stress analysis on a part and assembly

ADJUSTABLE SECTIONS VIEWS IN A PART FILE

In Chapter 6, you learned how to section an assembly to better see inside it. The same functionality is available while working in a part file. As in an assembly, there are four section commands available while working in a part file. To access the section commands, click on the View tab > Appearance panel as shown in the following image on the left. The following image in the middle shows a part before being sectioned and the view on the right shows the part sectioned with the Three Quarter Section View command. Note that the sections are temporary and sketches cannot be actively placed on a section. As always, you can create a sketch on a plane that was selected to create the section.

FIGURE 7.1

To section a part, follow these steps.

1. Select the desired section command on the View tab > Appearance panel.
2. Select any existing planar face(s), work plane(s) or origin plane(s).
3. Adjust the section depth by clicking and dragging the visible arrow in the graphics window; enter a value in the Offset cell in the graphics window as shown in the following image on the left. Another option to control the offset is to move the cursor over the Offset cell and scroll the wheel on the mouse. You can control the distance that each step takes when you scroll the wheel by right-clicking and click Virtual Movement > Scroll Step Size.
4. To accept the current location click the green check mark as shown in the following image in the middle, press the ENTER key or right-click and click OK in the menu.
5. If you are creating a quarter or three quarter section view, follow the same process to select a second plane.
6. While still in the command, you can specify which side of the section remains visible by right clicking and click the Flip Section or change the section type as shown in the following image on the right.
7. To finish the command, right-click and click Done from the menu.
8. To clear the section, click the End Section View command on the View tab > Appearance panel.

FIGURE 7.2

DESIGN VIEW REPRESENTATION IN A PART FILE

While working on a part file, you can save a design view representation that stores information related to the current viewpoint, part color, and work feature visibility. You can save as many design view representations as needed, but only one can be current. By default, there is a design view representation named Master, this master design view cannot be changed, deleted, or locked. To create a design view representation in a part file, follow these steps.

1. In the browser, right-click on View: Master and click New from the menu as shown in the following image on the left.
2. A generic design view will be created; you can rename the design view by slowly double-clicking on its name and type in a new name.
3. Once a design view is created, change the view point, part color, and work feature visibility and those changes are captured in the current design view.
4. Since these changes are actively captured to the current design view, you may want to lock a design view so any change to the current viewpoint, part color, or work feature visibility are not saved to the current design view. To lock a design view, right-click on its name in the browser and click Lock from the menu as shown in the following image on the right.
5. Create as many design views as needed.
6. Lock and unlock the current design view to capture the settings as needed.
7. To make a design view current, in the browser double-click on its name or right-click on its name and click Activate from the menu. The last settings in the design view or the locked settings in the design view will be displayed in the graphics window.

FIGURE 7.3

> Design View Representations can also be created in an assembly.

NOTE

EMBOSSED TEXT AND CLOSED PROFILES

To better define a part, you may need to have a shape or text either embossed (raised) or engraved (cut) into a model. In this section, you will learn how you can emboss or engrave a closed shape or text onto a planar or curved face. You can define a shape using the sketch commands on the 2D Sketch panel bar. The shape needs to be closed. There are two steps to emboss or engrave: first create the shape or text. Second, you emboss the shape or text onto the part. The text can be oriented in a rectangle or about an arc, circle, or line. The following sections describe the steps.

Step 1: Creating Rectangular Text

To place rectangular text onto a sketch, follow these steps:

1. Create a sketch.
2. Click the Text command on the Sketch tab > Draw panel, as shown in the following image on the left.
3. Pick a point or drag a rectangle where you will place the text. If a single point is selected, the text will fit on a single line and can be grip edited to resize the bounding box. If a rectangle is used, it defines the width for text wrapping and can be modified. The Format Text dialog box will appear.
4. In the Format Text dialog box, specify the text font and format style, and enter the text to place on the sketch. The following image shows the dialog box with text entered in the bottom pane.

FIGURE 7.4

5. When done typing text, click OK.
6. When the text is placed, a rectangular set of construction lines defines the perimeter of the text. You can click and drag a point on the rectangle and change the size of the rectangle. Dimensions or constraints can be added to these construction lines to refine the text's location and orientation, as shown in the following image.
7. To edit the text, move the cursor over the text in the graphics window, right-click, and select Edit Text from the menu. The same Format Text dialog box that was used to create the text will appear.

FIGURE 7.5

Step 2: Creating Text about Geometry

To place text about an arc, circle, or line in a sketch, follow these steps:

1. Make a sketch active.
2. Sketch and constrain an arc, circle, or line that the text will follow; it is recommended to make this geometry construction geometry as it will not be used as the geometry for the profile.
3. Click the Geometry Text command on the Sketch tab > Draw panel > Geometry Text command as shown in the following image on the left. The command may be under the Text command.
4. In the sketch, select the arc, circle, or line and the Geometry-Text dialog box will appear.

5. In the Geometry-Text dialog box, specify the text font and format style, direction, position, and start angle and enter the text to place on the sketch. The start angle is relative to the left quadrant point of a circle or the start point of the arc. The following image on the right shows the dialog box with text entered in the bottom pane.

FIGURE 7.6

6. When done entering text, click the Update button in the dialog box to see the preview in the graphics window.
7. Click OK to complete the operation. The following image shows text placed along an arc.
8. To edit the text, move the cursor over the text in the graphics window, right-click, and click Edit Geometry Text from the menu. The same Format Text dialog box that was used to create the text will appear.

FIGURE 7.7

Step 3: Embossing Text

To emboss text or a closed profile, click the Emboss command on the Model tab > Create panel, as shown in the following image on the left. The Emboss dialog box will appear, as shown in the following image on the right.

FIGURE 7.8

The Emboss dialog has the following options:

	Option	Description
	Profile	Select a profile, meaning closed shape or text, to emboss. You may need to use the Select Other command to select the text.
	Depth	Enter an offset depth to emboss or engrave the profile if you selected the Emboss from Face or Engrave from Face types.
	Top Face Color	Select a color from the drop-down list to define the color of the top face of the embossed area, not its lateral sides.
	Emboss from Face	Select this option to add material to the part.
	Engrave from Face	Select this option to remove material to the part.
	Emboss/ Engrave from Plane	Select this option to add and remove material from the part by extruding both directions from the sketch plane. Direction changes at the tangent point of the profile to a curved face.
	Flip Direction	Select either of these buttons to define the direction of the feature.
	Wrap to Face	Check this box for Emboss from Face or Engrave from Face types to wrap the profile onto a curved face. Only a single cylindrical or conical face can be selected. The profile will be slightly distorted as it is projected onto the face. The wrap stops if a perpendicular face is encountered.

To emboss a closed shape or text, follow these steps:

1. Click the Emboss command on the Create panel.
2. Define the profile by selecting a closed shape or text. If needed, use the Select Other command.
3. Select the type of emboss: Emboss from Face, Engrave from Face, or Emboss/Engrave from Plane.
4. Specify the depth, color, direction, and face as needed. If you selected Emboss/Engrave from plane option, you can also add a taper angle to the created emboss/engrave feature.

You edit the embossed feature like any other feature.

NOTE

EXERCISE 7-1: CREATING TEXT AND EMBOSS FEATURES

In this exercise, you emboss and engrave sketched profile objects on faces of a razor handle model. You then create a sketch text object and engrave it on the handle.

1. Open *ESS_E07_01.ipt* in the Chapter 07 folder.
2. Use the Free Orbit and Zoom commands to examine the part.
3. From the browser, turn on visibility of Sketch11 by right-clicking on Sketch11 and click Visibility from the menu and rotate the view to display the top triangular face, as shown in the following image.

FIGURE 7.9

4. Click the Emboss command on the Create panel to engrave a closed sketch profile. Notice that the only visible closed profile is automatically selected.
   - In the Emboss dialog box, click the Engrave from Face option (middle button).
   - Change the Depth to **.03125 in**.
   - Click the Top Face Color button, and choose Aluminum (Flat) from the Color dialog box drop-down list, as shown in the following image on the left.
5. Click OK twice to close both dialog boxes and create the engraved feature, as shown in the following image on the right.

FIGURE 7.10

6. Change to the Home View, and from the browser turn on the visibility of Sketch10.

7. Click the Emboss command from the Create panel.

   - For the profile, click the oblong and the six closed profiles. Be sure to select the left and right halves of the herringbone profiles, as shown in the following image on the left.
   - In the Emboss dialog box, click the Emboss from Face option (left button).
   - Change the Depth to **.015625 in**.
   - Click the Top Face Color button, and choose Black from the Color dialog box drop-down list, as shown in the following image on the right.

FIGURE 7.11

8. Click OK twice to close both dialog boxes and create the emboss feature.

9. Use the Orbit command to examine the engraved and embossed features.
10. Next you create a text sketch object. From the browser, turn on the visibility of Sketch9.
11. In the browser, double-click Sketch9 to make it the active sketch.
12. Click the Text command in the Draw panel.

   - Click in an open area below the part and under the left edge of the construction rectangle in the sketch to specify the insertion point and display the Format Text dialog box.
   - Click the Center and Middle Justification buttons.
   - Click the Italic option.
   - Change the % Stretch value to **120**.
   - Click in the text field, and type The SHARP EDGE.
   - In the text field, double-click on the word "The," and change the text size from 1.20 in to **.09 in**, as shown in the following image.

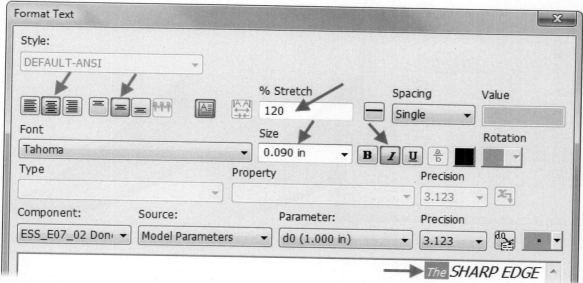

FIGURE 7.12

13. Click OK to create the text.
14. Draw a diagonal construction line coincident with the text bounding box corners, as shown in the following image.

The SHARP EDGE

FIGURE 7.13

15. Place a coincident constraint between the midpoint of the diagonal construction line of the text and Sketch9, as shown in the following image on the left. The completed operation is shown in the following image on the right.

FIGURE 7.14

16. Click the Finish Sketch command to finish editing the sketch.
17. Click the Emboss command.
    - For the profile, select the text object.
    - In the Emboss dialog box, click the Engrave from Face option.
    - Change the Depth to **.015625 in**.
    - For the direction, click the right button to change the direction down.
    - Click the Top Face Color button, and click Black from the Color drop-down list as shown in the following image on the left.
18. Click OK twice to close both dialog boxes and create the emboss feature.
19. Use the Free Orbit command to examine the engraved text feature as shown in the following image on the right.

FIGURE 7.15

20. Delete the Emboss3 feature you just created.
21. Edit Sketch9 by double-clicking on it in the browser.
22. In the sketch, delete the existing text, rectangle and angled lines; do NOT delete the yellow horizontal line in the middle of the part (your color may be different depending upon your color scheme).
23. Sketch and dimension a construction circle as shown.

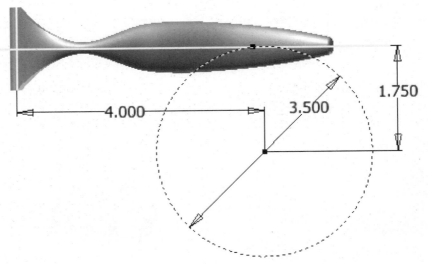

FIGURE 7.16

24. Click the Geometry Text command and select the circle for the Geometry selection.

    - Click the Center Justification button (middle button).
    - Change the Start Angle to **90.00 deg**.
    - Check the Bold button.
    - Change the text height to **.125 in**.
    - Click in the text field, and enter **The SHARP EDGE** as shown in the following image on the left (do not press ENTER, the words in your dialog box may be on a single line).
    - To preview the text on the geometry, click the Update button in the Geometry-Text dialog box as shown in the following image on the right.

FIGURE 7.17

    - Click OK to complete the command.

25. Click the Finish Sketch command to finish editing the sketch, and then change to the Home View.
26. Click the Emboss command.
    - For the profile, select the text object.
    - In the Emboss dialog box, click the Engrave from Face option.
    - Flip the direction so it points down.
    - Change the Depth to **.015625 in**.
    - Click the Top Face Color button, and click Black from the Color drop-down list.
27. Click OK twice to close both dialog boxes and create the emboss feature.

FIGURE 7.18

28. Use the Free Orbit command to examine the emboss features as shown in the following image.

FIGURE 7.19

29. Close the file. Do not save changes. End of exercise.

DIMENSION DISPLAY, RELATIONSHIPS, AND EQUATIONS

When creating part features, you may want to set up relationships between features and/or sketch dimensions. The length of a part may need to be twice that of its width, for example, or a hole may always need to be in the middle of the part. In Autodesk Inventor, you can use several different methods to set up relationships between dimensions. The following sections will cover these methods.

Dimension Display

When you create a dimension, it is automatically tagged with a label, or parameter name, that starts with the letter "d" and a number: for example, "d0" or "d27." The first dimension created for each part is given the label "d0." Each dimension that you place for subsequent part sketches and features is sequenced incrementally, one number at a time. If you erase a dimension, the next dimension does not go back and reuse the erased value. Instead, it keeps sequencing from the last value on the last dimension created. When creating dimensional relationships, you may want to view a dimension's display style to see the underlying parameter label of the dimension. Five options for displaying a dimension's display style are available.

Display as Value. Use to display the actual value of the dimensions on the screen.

Display as Name. Use to display the dimensions on the screen as the parameter name: for example, d12 or Length.

Display as Expression. Use to display the dimensions on the screen in the format of parameter# = value, showing each actual value: for example, d7 = 20 mm or Length = 50 mm.

Display as Tolerance. Use to display the dimensions on the screen that have a tolerance style for example, 40 ± .3.

Display as Precise Value. Use to display the dimensions on the screen and ignore any precision settings that are specified: for example, 40.3563123344.

To change the dimension display style, right-click in the graphics window. Click Dimension Display, as shown in the following image, or click the Tools tab > Document Settings from the Options panel and, on the Units tab, change the display style. After you select a dimension display style, all visible dimensions will change to that style. As you create dimensions, they will reflect the current dimension display style.

FIGURE 7.20

Dimension Relationships

Setting a dimensional relationship between two dimensions requires setting a relationship between the dimension you are creating and an existing dimension. When entering text in the Edit Dimension dialog box, enter the dimensions parameter name (d#) of the other dimension, or click the dimension with which you want to set the relationship in the graphics window. The following image on the left shows the Edit Dimension dialog box after selecting the 10 unit dimension to which the new dimension will be related. After establishing a relationship to another dimension, the dimension will have a prefix of fx:, as shown in the following image on the right.

FIGURE 7.21

Equations

You can also use equations whenever a value is required: examples include (d9/4)*2 or 50 in + 19 in. When creating equations, Autodesk Inventor allows prefixes, precedence, operators, functions, syntax, and units. To see a complete listing of valid options, use the Help system, and search for Functions, Prefixes, and Algebraic Operators.

You can enter numbers with or without units; when no unit is entered, the default unit will be assumed. As you enter an equation, Autodesk Inventor evaluates it. An invalid expression will appear in red, and a valid expression will appear in black. For best results while using equations, include units for every term and factor in the equation.

To create an equation in any field, follow these steps:

1. Click in the field.
2. Enter any valid combination of numbers, parameters, operators, or built-in functions. The following image shows an example of an equation that uses both millimeters and inches for the units.
3. Press ENTER or click the green checkmark to accept the expression.

NOTE Use ul (unitless) where a number does not have a unit. For example, use a unitless number when dividing, multiplying, or specifying values for a pattern count.

FIGURE 7.22

PARAMETERS

Another way to set up relationships between dimensions is to use parameters. A parameter is a user-defined name that is assigned a numeric value, either explicitly or through equations. You can use multiple parameters in an equation, and you can use parameters to define another parameter such as *depth = length − width*. You can use a parameter anywhere a value is required. There are four types of parameters: model, user, reference, and linked.

> **NOTE** Add-ins such as Simulation: Stress Analysis and Design Accelerator can automatically add parameter groups. Depending on the Add-in, you may or may not be able to remove those groups.

Model. This type is created automatically and assigned a name when you create a sketch dimension, feature parameter such as an extrusion distance, draft angle, or coil pitch, and the offset, depth, or angle value of an assembly constraint. Autodesk Inventor assigns a default name to each Model Parameter as you create it. The default name format is a "d" followed by an integer incremented for each new parameter. You can rename Model Parameters via the Parameters dialog box or specify a name upon initial creation in the Edit Dimension dialog box. This method is called "creating a parameter on the fly" and can be done in any field that you can specify a parametric dimensional value. The parameter that is being defined by the field will be renamed to the name specified at the beginning of the equation. For example, if you were defining a rectangular sketch and wanted to define a parameter named Length while dimensioning the sketch, you would enter "Length=10" (without the quotation marks) in the Edit Dimension dialog box. A parameter with the name of Length would be automatically created and available.

User. This type is created manually from an entry in the Parameters dialog box. You can add a numeric parameter, or if using rules-based design, you can add text or true-false user parameters.

Reference. This type is created automatically when you create a driven dimension. Autodesk Inventor assigns a default name to each reference parameter as you create it. The default name format is a "d" followed by an integer incremented for each new parameter. You can rename reference parameters via the Parameters dialog box.

Linked

This type is created via a Microsoft Excel spreadsheet and is linked into a part or assembly file. This is done using the Parameters command on the Manage tab > Parameters panel, as shown in the following image.

FIGURE 7.23

After you click the Parameters command, the Parameters dialog box is displayed. The following image shows an example with some model, user, and reference parameters created. The Parameters dialog box is divided into two sections: Model Parameters and User Parameters. A third section called Reference Parameters is displayed if driven dimensions exist. The values from dimensions or assembly constraints used in the active document automatically fill the Model Parameters section. The User Parameters section is defined manually. You can change the names and equations of both types of parameters, and you can add comments by double-clicking in the cell and entering the new information. The column names for Model and User Parameters are the same, and the following sections define them.

Parameter Name. The name of the parameter will appear in this cell. To change the name of an existing parameter, click in the box, and enter a new name. When creating a new User Parameter, enter a new name after clicking the Add Numeric, Add Text, or Add True/False button. When you update the model, all dependent parameters update to reflect the new name.

Unit/Type. Enter a new unit of measurement for the parameter in this cell. With Autodesk Inventor, you can build equations that include parameters of any unit type. All length parameters are stored internally in centimeters; angular parameters are stored internally in radians. This becomes important when you combine parameters of different units in equations.

Equation. The equation will appear in this cell, and it will determine the value of the parameter. If the parameter is a discrete value, the value appears in rounded form to match the precision setting for the active document. To change the equation, click on the existing equation, and enter the new equation.

Nominal Value. The nominal tolerance result of the equation will appear in this cell, and it can only be modified by editing the equation.

Tol. (Tolerance). From the drop-down list, select a tolerance condition: upper, median, nominal, or lower.

Model Value. The actual calculated model value of the equation in full precision will appear in this cell. This value reflects the current tolerance condition of the parameter.

Key. Click to specify which defined parameters will be identified as a key parameter.

Export Parameters. Click to export the parameter to the Custom tab of the iProperties dialog box. The parameter will also be available when using the Derive command as well as in the bill of materials and Parts List Column Chooser dialog boxes.

Comment. You may choose to enter a comment for the parameter in this cell. Click in the cell, and enter the comment.

FIGURE 7.24

User Parameters

User Parameters are ones that you define in a part or an assembly file. Parameters defined in one environment are not directly accessible in the other environment. If parameters are to be used in both environments, use a linked parameter via a common Microsoft Excel spreadsheet. You can use a parameter any time a numeric value is required. When creating parameters, follow these guidelines:

- Assign meaningful names to parameters, as other designers may edit the file and will need to understand your thought process. You may want to use the comments field for further clarification.
- The parameter name cannot include spaces, mathematical symbols, or special characters. You can use these to define the equation.

- Autodesk Inventor detects capital letters and uses them as unique characters. Length, length, and LENGTH are three different parameter names.
- When entering a parameter name where a value is requested, the upper- and lowercase letters must match the parameter name.
- When defining a parameter equation, you cannot use the parameter name to define itself; for example, Length = Length/2 would be invalid.
- Duplicate parameter names are not allowed. Model, User, and Spreadsheet-driven parameters must have unique names.

To create and use a User Parameter, follow these steps:

1. Click the Parameters command on the Manage tab > Parameters panel.
2. Click the appropriate Add button for the type of parameter you want to use (Numeric, Text, True/False) at the bottom of the Parameters dialog box.
3. Enter the information for each of the cells.
4. After creating the parameter(s), you can enter the parameter name(s) anywhere that a value is required. When editing a dimension, click the arrow on the right, and select List Parameters from the menu, as shown in the following image. All the available User Parameters and any named parameters will appear in a list similar to that in the image on the right. Click the desired parameter from the list.

FIGURE 7.25

Linked Parameters

If you want to use the same parameter name and value for multiple parts, you can do so by creating a spreadsheet using Microsoft Excel. You then embed or link the spreadsheet into a part or assembly file through the Parameters dialog box. When you **embed** a Microsoft Excel spreadsheet, there is no link between the spreadsheet and the parameters in the Autodesk Inventor file, and any changes to the original spreadsheet will not be reflected in the Autodesk Inventor file. When you **link** a Microsoft Excel spreadsheet to an Autodesk Inventor part or assembly file, any changes in the original spreadsheet will update the parameters in the Autodesk Inventor file. You can link more than one spreadsheet to an Autodesk Inventor file, and you can link each spreadsheet to multiple part and assembly files. By linking a spreadsheet to an assembly file and to the part files that comprise the assembly, you can drive parameters from both environments from the same spreadsheet. There is no limit to the number of part or assembly files that can reference the same spreadsheet. Each linked spreadsheet appears in the browser under the 3rd Party folder.

When creating a Microsoft Excel spreadsheet with parameters, follow these guidelines:

- The data in the spreadsheet can start in any cell, but the cells must be specified when you link or embed the spreadsheet.
- The data can be in rows or columns, but they must be in this order: parameter name, value or equation, unit of measurement, and (if desired) a comment.
- The parameter name and value are required, but the other items are optional.
- The parameter name cannot include spaces, mathematical symbols, or special characters. You can use these to define the equation.
- Parameters in the spreadsheet must be in a continuous list. A blank row or column between parameter names eliminates all parameters after the break.
- If you do not specify a unit of measurement for a parameter, the default units for the document will be assigned when the parameter is used. To create a parameter without units, enter "ul" in the Units cell.
- Only those parameters defined on the first worksheet of the spreadsheet become linked to the Autodesk Inventor file.
- You can include column or row headings or other information in the spreadsheet, but they must be outside the block of cells that contains the parameter definitions.

The following image on the left shows three parameters that were created in rows with a name, equation, unit, and comment. The right side of the image shows the same parameters created in columns.

	A	B	C	D
1	Length	50	mm	Length of Plate
2	Width	Length/2	mm	Width of Plate
3	Depth	10	mm	Depth of Plate

	A	B	C
1	Length	Width	Depth
2	50	Length/2	10
3	mm	mm	mm
4	Length of Plate	Width of Plate	Depth of Plate

FIGURE 7.26

After you have created and saved the spreadsheet, you can create parameters from it by following these steps:

1. Click the Parameters command on the Manage tab > Parameters panel.
2. Click Link on the bottom of the Parameters dialog box.
3. The Open dialog box, similar to that shown in the following image, will appear.

FIGURE 7.27

4. Navigate to and select the Microsoft Excel file to use.
5. In the lower-left corner of the Open dialog box, enter the start cell for the parameter data.
6. Select whether the spreadsheet will be linked or embedded.
7. Click the Open button: a section showing the parameters is added to the Parameters dialog box, as shown in the following image. If you embedded the spreadsheet, the new section will be titled Embedding #.
8. To complete the operation, click Done.

C:\Essentials Plus								
BaseExtrusion	mm	30 mm	30.000000	○	30.000000	☐	☐	Extrusion distance for the base feature
Draft	deg	5 deg	5.000000	○	5.000000	☐	☐	Draft for all features

FIGURE 7.28

To edit the parameters that are linked, follow these steps:

1. Open the Microsoft Excel file.
2. Make the required changes.
3. Save the Microsoft Excel file.
4. Open the Autodesk Inventor part or assembly file that uses the spreadsheet.

You can also follow these steps when a spreadsheet has been linked. If you chose to embed the spreadsheet, the following steps must be used to edit the parameters:

1. Open the Autodesk Inventor part or assembly file that uses the spreadsheet.
2. Expand the 3rd Party folder in the browser.

3. Either double-click the spreadsheet or right-click and select Edit from the menu, as shown in the following image.

FIGURE 7.29

4. The Microsoft Excel spreadsheet will open in a new window for editing.
5. Make the required changes.
6. Save the Microsoft Excel file.
7. Activate the Autodesk Inventor part or assembly file that uses the spreadsheet.
8. Click the Update tool from the standard toolbar.

NOTE: If you embedded the spreadsheet, the changes will not be saved back to the original file but will only be saved internally to the Autodesk Inventor file.

EXERCISE 7-2: RELATIONSHIPS AND PARAMETERS

In this exercise, you create a sketch and dimension it. You then set up relationships between dimensions, and define User Parameters in both the model and an external spreadsheet.

1. Start a new file based on the default *Standard (in).ipt* file.
2. Draw the geometry, as shown in the following image. The lower-left corner should be located at the origin point, and the arc is tangent to both lines. The size should be roughly 25 in wide by 40 in high.

FIGURE 7.30

3. Add the four dimensions, as shown in the following image.

FIGURE 7.31

4. Double-click the 10 in radius dimension of the arc, and for its value, select the 25 in horizontal dimension, as shown in the following image.

FIGURE 7.32

5. In the Edit Dimension dialog box, type **/4** after the d#, creating "d0/4." The 0 may be a different number depending upon the order in which your geometry was dimensioned. Click the green checkmark to accept the dimension value in the Edit Dimension dialog box.
6. Double-click the 10 in vertical dimension, and for its value, select the 30 in vertical dimension. Type **/2** after the d#. Click the green checkmark to accept the dimension value in the Edit Dimension dialog box.
7. Change the value of the 25 in horizontal dimension to **50 in** and the 30 in vertical dimension to **40 in**. When done, your sketch should resemble the following image. The fx: text that is displayed before the two dimensions denotes that they are equation driven or have a reference to a parameter.

FIGURE 7.33

8. Next create parameters and drive the sketch from them. Click the Parameters command on the Manage tab > Parameters panel.
9. In the Parameter dialog box, create a User Parameter by clicking the Add Numeric button. Type in the following information:
   - Parameter Name = **Length**
   - Units = **in**
   - Equation = **75**
   - Comment = **Bottom length**
10. Create another User Parameter by selecting the Add Numeric button, and type in the following information:
    - Parameter Name = **Height**
    - Units = **in**
    - Equation = **Length/2**
    - Comment = **Height is half of the length**

   When done, the User Parameter area should resemble the following image.

User Parameters								
Length	in	75 in	75.000000	○	75.000000	☐	☐	Bottom length
Height	in	Length / 2 ul	37.500000	○	37.500000	☐	☐	Height is half of the length

FIGURE 7.34

11. Close the Parameter dialog box by clicking Done.
12. Change the dimension display by right-clicking the graphics window, and then click Dimension Display > Expression.
13. Double-click the bottom horizontal dimension, and change its value to the parameter Length as follows: click the arrow; on the right side of the menu, click List Parameters, click Length from the list, and then click the checkmark in the dialog box.
14. Double-click the right vertical dimension, and change its value to Height. When done, your screen should resemble the following image.

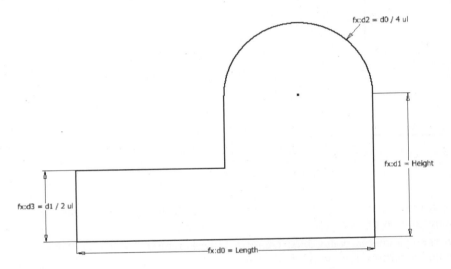

FIGURE 7.35

15. Click the Parameters command on the Manage tab > Parameters panel. From the User Parameters section, click on the equation cell of the parameter name Length, and change its value to **35**.
16. Click in the equation cell of the parameter name Height; change its value to **50** and the comment field to "**Height is half of the right side.**" When done making the changes, the User Parameter area should resemble the following image.

User Parameters								
Length	in	35 in	35.000000	○	35.000000	☐	☐	Bottom length
Height	in	50 in	50.000000	○	50.000000	☐	☐	Height is half of the right side

FIGURE 7.36

17. To complete the changes, click the Done button.

18. Click the Finish Sketch command on Sketch tab > Exit panel, and the sketch should update to reflect the changes, as shown in the following image.

FIGURE 7.37

19. Save the file as ESS_E07_02.ipt.
20. Now create a spreadsheet that has two parameters. Create a Microsoft Excel spreadsheet with the column names Parameter Name, Equation, Unit, and Comment, as shown in the following image. Enter the following data:

    - Parameter Name = **BaseExtrusion**
    - Equation = **30**
    - Units = **in**
    - Comment = **Extrusion distance for the base feature**
    - Parameter Name = **Draft**
    - Equation = **5**
    - Units = **deg**
    - Comment = **Draft for all features**

	A	B	C	D
1	Parameter Name	Equation	Unit	Comment
2	BaseExtrusion	30	in	Extrusion distance for the base feature
3	Draft	5	deg	Draft for all features

FIGURE 7.38

21. Save the spreadsheet as ESS_07_Parameters.xlsx. in the Chapter 07 folder.
22. Make Autodesk Inventor the current application, and then click the Parameters command.
23. In the Parameter dialog box, click the Link button and select the ESS_07_Parameters.xlsx file in the Chapter 07 folder, but do not click Open yet.
24. For the Start Cell, enter **A2**. If you fail to do this, no parameters will be found.
25. Click the Open button, and a new spreadsheet area will appear in the Parameters dialog box, as shown in the following image. Click the Done button to complete the operation.

C:\Essentials Plus...							
BaseExtrusion	in	30 in	30.000000	○	30.000000	☐ ☐	Extrusion distance for the base feature
Draft	deg	5 deg	5.000000	○	5.000000	☐ ☐	Draft for all features

FIGURE 7.39

26. Change to the home view.
27. Click the Extrude command from the Model tab > Create panel, and click the More tab in the Extrude dialog box, as shown in the following image. For the Taper's value, enter **Draft** or click the arrow from the menu, click List Parameters, and then click the parameter Draft.

FIGURE 7.40

28. Click the Shape tab. For the Distance value, use the parameter BaseExtrusion, as shown in the following image.

FIGURE 7.41

29. To complete the operation, click the OK button. Your model should resemble the following image.

FIGURE 7.42

30. In the browser, expand the 3rd Party icon. Either double-click or right-click on the name *ESS_08_Parameters.xls,* and select Edit from the menu.
31. In the Microsoft Excel spreadsheet, make the following changes: For the Parameter Name "BaseExtrusion," change the Equation to **50**. For the Parameter Name "Draft," change the Equation to **–3**.
32. Save the spreadsheet, and close Microsoft Excel.
33. Make Autodesk Inventor the current application, and click the Local Update command from the Quick Access Bar. When done, your model should resemble the following image.

FIGURE 7.43

34. Close the file. Do not save changes. End of exercise.

SWEEP FEATURES

Unlike other sketched features, a sweep feature requires two unconsumed sketches: a profile to be swept and a path that the profile will follow. Additional profiles can be used as guide rails or a surface can also be used to help shape the feature. The profile sketch and the path sketch cannot lie on the same or parallel planes. The path can be an irregular shape or can be based on a part edge by projecting and including the edges onto the active sketch. The path can be either an open or closed profile and can lie in a plane or lie in multiple planes (3D Sketch). Handles, cabling, and piping are examples of sweep features. A sweep feature can be a base or a secondary feature. To create a sweep feature, use the Sweep command on the Model tab > Create panel, as shown in the following image on the left. The Sweep dialog will appear, as shown in the following image on the right. The following list includes descriptions of the options in the Sweep dialog box.

FIGURE 7.44

Profile

Click this button to select the sketch profile to sweep. If the Profile button is depressed and red, it means that you need to select a sketch or sketch area. If there are multiple closed profiles, you will need to select the profile that you want to sweep. If there is only one possible profile, Autodesk Inventor will select it for you and you can skip this step. If you selected the wrong profile or sketch area, depress the Profile button, and deselect the incorrect sketch by clicking it while holding down the CTRL key. Release the CTRL key, and select the desired sketch profile.

Path

Click this button to select the path along which to sweep the profile. The path can be an open or a closed profile but must pierce the sketch that the profile was drawn on. You can also use edge(s) of a part as a path. The profile is typically perpendicular to and intersects with the start point of the path. The start point of the path is often projected into the profile sketch to provide a reference point.

FIGURE 7.45

Solids

If there are multiple solid bodies, click this button to choose the solid body(ies) to participate in the operation.

Output Buttons

In the Output section, click Solid to create a solid feature or Surface to create the feature as a surface.

Operation Buttons

This is the column of buttons down the middle of the dialog box. By default, the Join operation is selected. Use the operation buttons to add or remove material from the part using the Join or Cut options or to keep what is common between the existing part and the completed sweep using the Intersect option.

- *Join:* Adds material to the part.
- *Cut:* Removes material from the part.
- *Intersect:* Creates a new feature from the shared volume of the sweep feature and existing part volume. Material not included in the shared volume is deleted.

Type Box
Path

Creates a sweep feature by sweeping a profile along a path.

FIGURE 7.46

Orientation
Path

Holds the swept profile constant to the sweep path. All sweep sections maintain the original profile relationship to the path, as shown in the previous image.

Parallel

Holds the swept profile parallel to the original profile as shown in the previous image.

Taper Enter a value for the angle you want the profile to be drafted. By default, the taper angle is 0, as shown in the previous image.

Path & Guide Rail
Creates a sweep feature by sweeping a profile along a path and uses a guide rail to control scale and twist of the swept profile. The following image shows the options for the path and guide rail type.

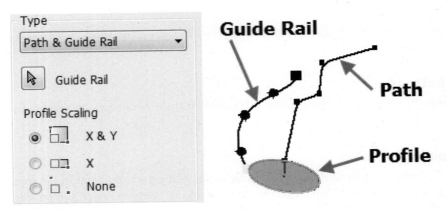

FIGURE 7.47

Guide Rail. Select a guide curve or rail that controls the scaling and twist of the swept profile. The guide rail must touch the profile plane. If you project an edge to position the guide rail and the projected edge is not the path, change the projected edge to a construction line.

Profile Scaling. Specify how the swept section scales to meet the guide rail. The following image shows the path and the profile and guide rail from the previous image with the three different profile scaling options.

X and Y. Scales the profile in both the X and Y directions as the sweep progresses.

X. Scales the profile in the X direction as the sweep progresses. The profile is not scaled in the Y direction.

None. Keeps the profile at a constant shape and size as the sweep progresses. Using this option, the rail controls only the profile twist and is not scaled in the X or Y direction.

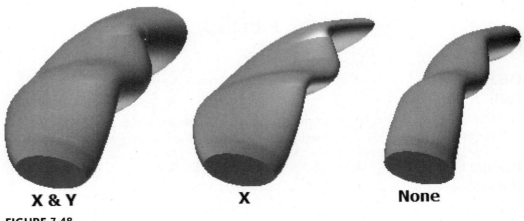

FIGURE 7.48

Path & Guide Surface

Creates a sweep feature by sweeping a profile along a path and a guide surface. The guide surface controls the twist of the swept profile. For best results, the path should touch or be near the guide surface.

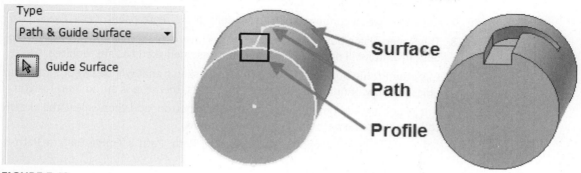

FIGURE 7.49

The following image on the left shows a sweep with just the path. Notice that the back inside edge of the cut is angled from the surface. The following image on the right shows the same sweep with a guide surface. Notice that the back inside edge is parallel to the top of the surface.

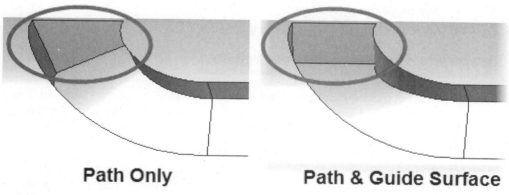

FIGURE 7.50

Optimize for Single Selection

When this option is checked, Autodesk Inventor automatically advances to the next selection option once a single selection is made. Clear the checkbox when multiple selections are required.

Preview

The button with the eyeglasses provides a solid preview of the sweep based on the current selections. If Preview is enabled and no preview appears in the graphics window, then the sweep feature will not be created.

Creating a Sweep Feature

In this section, you will learn how to create sweep features. You first need to have two unconsumed sketches. One sketch will be swept along the second sketch that represents the path. To create a sweep feature, follow these steps:

1. Create two unconsumed sketches: one for the profile and the other for the path. If you are going to use existing part edges for the path you don't need to create

a sketch for the path. The profile and path must lie on separate nonparallel planes. It is recommended that the profile intersect the path. Use work planes to place the location of the sketches, if required. The sketch that you use for the path can be open or closed. Add dimensions and constraints to both sketches as needed. If needed, create a sketch that will be used as a rail, or create a surface to be used as a guide surface.

2. Click the Sweep command on the Model tab > Create panel.
3. The Sweep dialog box will appear. If two unconsumed sketches do not exist, Autodesk Inventor will notify you that two unconsumed sketches are required.
4. Click the Profile button, and then select the sketch that will be swept in the graphics window. If only one closed profile exists, this step is automated for you.
5. If it is not already depressed, click the Path button, and then select the sketch to be used as the path in the graphics window.
6. Select the type of sweep you are creating: Path, Path & Guide Rail, or Path & Guide Surface.
7. Define whether or not the resulting sweep will create a solid or a surface by clicking either the Solid or the Surface button in the Output area. Select the required options, as outlined in the previous descriptions.
8. If this is a secondary feature, click the operation that will specify whether material will be added or removed or if what is common between the existing part and the new sweep feature will be kept.
9. If you want the sweep feature to have a taper, click on the More tab, and enter a value for the taper angle.
10. Click the OK button to complete the operation.

EXERCISE 7-3: CREATING SWEEP FEATURES

In the first part of this exercise, you create a component with the sweep command. Three sketches will be created, one each for the profile, the path, and the guide rail. In the second part of the exercise, you sweep a profile using edges on a part as the path.

1. Open *ESS_E07_03_1.ipt* from the Chapter 07 folder.
2. Click the Sweep command on the Create panel. The profile and path are automatically selected since only one closed profile (ellipse) and one open sketch (arc) are visible. Note that the profile or path can be open or closed, but if it is closed, you will need to manually select it; if the profile is open, a surface will be created.
3. Change the Orientation from Path to Parallel, as shown in the following image. Notice how the profile changes orientation.

FIGURE 7.51

4. Click OK to create the sweep feature.
5. Turn on the visibility of the guide rail by moving the cursor over the entry Sketch Guide Rail in the browser. Right-click, and click Visibility from the menu.
6. Rotate the viewpoint, and notice that the front end of the guide rail is coincident to the profile. To better see the geometry, change the Visual Style to Wireframe from the Navigation Bar or from the View tab > Appearance panel > Visual Style.
7. Right-click in the graphics window, and click Home View from the menu.
8. Edit the sweep feature by moving the cursor over the "Sweep1" entry in the browser and double-click.
9. Change the type to Path & Guide Rail and, in the graphics window, select the spline for the Guide Rail, as shown in the following image.

FIGURE 7.52

10. Notice that the Profile Scaling is set to X & Y. Click OK to create the part. Note that the sweep could have been completed as a Path & Guide Rail immediately.
11. Change the Visual Style back to Shaded. The following image on the left and in the middle shows the Profile Scaling set to X & Y.
12. Rotate the model, and notice that the profile stretches in both X and Y directions.
13. Right-click in the graphics window, and click Home View from the menu.
14. Edit the sweep feature by clicking on the model in the graphics window and click the Edit Sweep button from the mini-toolbar that is displayed in the graphics window.
15. Change the Profile Scaling to X, as shown in the following image on the right, and click OK to update the sweep feature.
16. Rotate the viewpoint, and notice that the profile stretches only in the X direction.

FIGURE 7.53

17. Close the file.
18. Open *ESS_E07_03_2.ipt* in the Chapter 07 folder.
19. Zoom in on the lower right-side of the part, as shown in the following image on the left.
20. Click on the front face of the part and click Create Sketch from the mini-toolbar as shown in the following image on the left.
21. The viewpoint will change so you are looking directly at the sketch. However, it is difficult to see where you are sketching. To return to your previous view, press the F5 key.
22. Sketch a circle on the projected point and add a **.0625 in** dimension, as shown in the following image on the right.

FIGURE 7.54

23. Click the Finish Sketch command on the Exit panel.
24. Next, you sweep the circle along edges of the part using the Sweep command. Click the Sweep command on the Create panel.
25. For the profile select inside the circle and the Path button will then be active. Click on a top-outside edge of the part, Inventor will automatically select all of the top-outside edges that go around the part, as shown in the following image on the left.
26. In the Sweep dialog box, click the Cut operation to remove material.
27. Click the OK button in the Sweep dialog box. The following image on the right shows the completed part.

FIGURE 7.55

28. Use the Zoom and Free Orbit commands to examine the Sweep feature.
29. Close the file. Do not save changes. End of exercise.

3D Sketching

To create a sweep feature whose path does not lie on a single plane, you need to create a 3D sketch that will be used for the path. You can use a 3D sketch to define the path for a lip or to define the routing path for an assembly component, such as a pipe or duct work that crosses multiple faces on different planes. You need to define a 3D sketch in the part environment, and you can do this within an assembly or in its own part file.

3D Sketch Environment

The 3D sketch environment is used to create 3D or a combination of both 2D and 3D curves. Before creating a 3D sketch, change the environment to the 3D sketch environment by clicking the Create 3D Sketch command on the Model tab > Sketch panel beneath the Create 2D Sketch command, as shown in the following image on the left. The commands on the ribbon will change to the 3D Sketch commands, as shown in the following image on the right. The most common 3D sketch commands are explained throughout this chapter. While in the 3D Sketch environment, all features appear in the browser with a 3D sketch name. Once a feature uses the 3D sketch, it will be consumed under the new feature in the browser.

 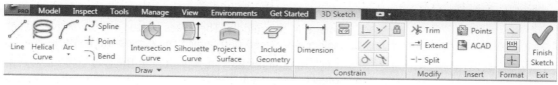

FIGURE 7.56

3D Sketch from Intersection Geometry

Another way to create a 3D path is to use geometry that intersects with the part. If the intersecting geometry defines the 3D path, you can use it. The intersection can be defined by a combination of any of the following: planar or nonplanar part faces, surface faces, a quilt, or work planes. To create a 3D path from an intersection, follow these steps:

1. Create the intersecting features that describe the desired path.
2. Change to the 3D Sketch environment by clicking the Create 3D Sketch command on the Sketch panel under the 2D Sketch command.
3. Click the Intersection Curve command, as shown in the following image on the left on the Draw panel.
4. The 3D Intersection Curve dialog box appears, as shown in the following image in the middle.
5. Select the two intersecting geometry.
6. To complete the operation, click OK.

The following image on the right shows a 3D path being created on a cylindrical face at the edge defined by a work plane that intersects it at an angle.

FIGURE 7.57

Project to Surface

While in a 3D sketch, you can project curves, 2D or 3D geometry, part edges, and points onto a face or onto selected faces of a surface or solid. To project curves onto a face, follow these steps:

1. Create the solid or surface onto which the curves will be projected.
2. Create the curves that will be projected onto the face(s).
3. Click the 3D Sketch command on the Sketch panel.
4. Click the Project to Surface command on the Draw panel, as shown in the following image on the left.
5. The Project Curve to Surface dialog box will appear, as shown in the following image in the middle. The Faces button will be active. In the graphics window, select the face(s) onto which the curves will be projected.
6. Click the Curves Button, and then in the graphics window, select the individual objects to project. The following image on the right shows the face and curves selected.

FIGURE 7.58

7. In the Output area, select one of the following options.

	Project along Vector	Specify the vector by clicking the Direction button and selecting a plane, edge, or axis. If a plane is selected, the vector will be normal (90°) from the plane. The curves will be projected in the direction of the vector.
	Project to Closest Point	Projects the curves onto the surface normal to the closest point.
	Wrap to Surface	The curves are wrapped around the curvature of the selected face or faces.

8. Click OK.

By default, the projected curves are linked to the original curve. If the original curves change size, they will be updated. To break the link, move the cursor into the browser over the name of the Projected to Surface entry, right-click, and select Break Link in the menu, as shown in the following image. You could also display the sketch constraints and delete the reference constraints. You can also change the way the curves were projected by right-clicking on the Project to Surface entry in the browser and click Edit Projection Curve from the menu as shown in the following image, and the Project Curve to Surface dialog box will appear. Make the changes as required.

FIGURE 7.59

PROJECT TO 3D SKETCH

Another method to project 2D geometry onto a nonplanar face is to use the Project to 3D Sketch command. The 3D Sketch command will project 2D sketch geometry onto a nonplanar face while automatically creating a 3D sketch. To use the Project to 3D Sketch command, follow these steps:

1. First create a 2D sketch and sketch and constrain geometry as needed; or make an existing sketch active that contains the geometry that you want to project.
2. Click the Project to 3D Sketch command from the Sketch tab > Draw panel as shown in the following image on the left.
3. In the Project to 3D Sketch dialog box, check the Project option.
4. Select a face or faces onto which the geometry will be projected. All of the geometry on the active sketch will preview how it will be projected on the selected face(s) as shown in the following image on the right.

FIGURE 7.60

5. Click OK to create a 3D sketch and project all the geometry on the active sketch to the selected face(s).

EXERCISE 7-4: 3D SKETCH—SWEEP FEATURES

In this exercise, you use the Include Geometry, 3D Sketch from Intersection Geometry, Project to Surface, Project to 3D Sketch and the Sweep commands.

1. In this portion of this exercise, you will use the Include Geometry command to create a path that the sweep command will use. Open *ESS_E07_04_1.ipt* from the Chapter 07 folder. The part contains a sketch that has a constrained circle.
2. Create a 3D Sketch by clicking the Create 3D Sketch command on the Sketch panel under the Create 2D Sketch command.
3. Click the Include Geometry command from the 3D Sketch tab > Draw panel.
4. Select the nine edges that go around the part as shown in the following image on the left.
5. Finish the 3D Sketch by pressing the ESC key or right-click and click Done from the marking menu.
6. Finish the 3D sketch by clicking on Finish Sketch from the Exit panel.

7. Use the sweep command with the following options:
    - Use the circle as the profile.
    - Use the included geometry as the path (use the Select Other command to select the Sweep Path as shown in the following image in the middle).
    - Change the operation to cut.
    - Click OK to create the sweep. When done, your screen should resemble the following image on the right.

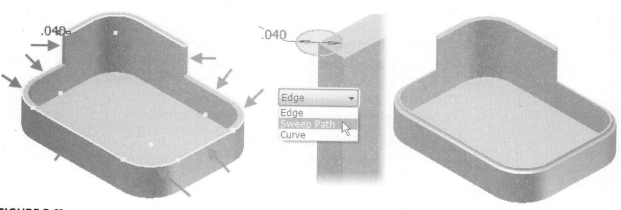

FIGURE 7.61

Note that you could have used the Sweep command without creating a 3D sketch and just select the edges on the part for the path.

8. Close the file. Do not save changes.
9. In this portion of this exercise, you will use the 3D Sketch from Intersection Geometry command to create a path that the sweep command will use. Open *ESS_E07_04_2.ipt* from the Chapter 07 folder. The part contains an angled work plane.
10. Create a 3D Sketch by clicking the Create 3D Sketch command on the Sketch panel under the Create 2D Sketch command.
11. Click the Intersection Curve command from the 3D Sketch tab > Draw panel.
12. Select the work plane and the circular face of the part as shown in the following image on the left.
13. Click OK to create the curve.
14. Turn off the visibility of the work plane and, when done, your screen should resemble the following image on the right.

FIGURE 7.62

15. Finish the 3D sketch by clicking on Finish Sketch from the Exit panel.
16. Next you create a 2D sketch that the sweep command will use as the profile. Click the Create 2D Sketch command on the Sketch panel under the Create 3D Sketch command, and from the Origin folder in the browser click the YZ Plane, as shown in the following image on the left.
17. Change to the Home view.
18. Press the F7 key to slice the graphics.
19. Use the Project Geometry command to project the geometry that was created from the 3D Sketch from Intersection Geometry command.
20. Change the projected line to a construction line. This will prevent the projected line from being used when you create the sweep.
21. Create a **.5 in** diameter circle on the bottom left corner of the projected geometry as shown in the following image in the middle.
22. Finish the Sketch.
23. Use the sweep command with the following options:
    - Use the circle as the profile.
    - Use the geometry that was created from the intersecting plane and face for the sweep path.
    - Change the operation to cut.
    - Click OK to create the sweep. When done, your screen should resemble the following image on the right.

FIGURE 7.63

24. Close the file. Do not save changes.
25. In this portion of this exercise, you will use the Project to 3D Sketch command to project geometry onto a cylindrical face. Open *ESS_E07_04_3.ipt* from the Chapter 07 folder. The part contains a constrained sketch.
26. Make Sketch2 active by double-clicking on Sketch2 in the browser.
27. To better see the part, change to the Home view.
28. Click the Project to 3D Sketch command on the Sketch tab > Draw panel.
29. In the Project to 3D Sketch dialog box, check the Project option and click the inside circular face on the part as shown in the following image on the left.
30. Click OK to complete the operation.
31. Finish the sketch.
32. Turn off the visibility of Sketch2. When done, your screen should resemble the following image on the right.

FIGURE 7.64

33. If desired, you can create a profile and sweep it along the projected geometry as you did earlier in this exercise. Note that you could have also used the Project to Surface command in the 3D sketch tab to project the geometry onto the face.
34. Close the file. Do not save changes. End of exercise.

LOFT FEATURES

The Loft command creates a feature that blends a shape between two or more different sections or profiles. A point can also be used to define the beginning and ending section of the loft. Loft features are used frequently when creating plastic or molded parts. A loft is similar to a sweep, but it can have multiple sections and rails. Many of these types of parts have complex shapes that would be difficult to create using standard modeling techniques. You can create loft features that blend between two or more cross-section profiles that reside on different planes. You can also control the area of a specific section in the loft. You can use a rail, multiple rails, or a centerline to define a path(s) that the loft will follow. There is no limit to the number of sections or rails that you can include in a loft feature. Four types of geometry are used to create a loft: sections, rails, centerlines, and points. The following sections describe these types of geometry.

Create a Loft

To create a loft feature, follow these steps:

1. Create the profiles or points that will be used as the sections of the loft. If required, use work features, sketches, or projected geometry to position the profiles.
2. Create rails or a centerline that will be used to define the direction or control of the shape between sections.
3. Click the Loft command on the Create panel, as shown in the following image on the left.
4. On the Curves tab of the Loft dialog box, the Sections option will be the default. In the graphics window, click the sketches, face loops, or points in the order in which the loft sections will blend.
5. If rails are to be used in the loft, click Click to add in the Rails section of the Curves tab, and then click the rail or rails.
6. If needed, change the options for the loft on the Conditions and Transition tabs.

NOTE The loft options are also available by right-clicking in a blank area in the graphics window and clicking an option on the menu.

FIGURE 7.65

The following image shows a loft created from two sections and a centerline rail. The image on the left shows two sections and a centerline in their sketch stages, shown in the top view. The image on the right shows the completed loft.

FIGURE 7.66

The following sections explain the options in the Loft dialog box and on its tabs.

Curves Tab

The Curves tab, as shown in the first image in this section, allows you to select which sketches, part edges, part faces, or points will be used as sections to select whether or not a rail or centerline will be used and to determine the output condition.

Sections

You can define the shape(s) between which the loft will blend. The following rules apply to sections:

1. There is no limit to the number of sections that you can include in the loft feature.
2. Sections do not have to be sketched on parallel planes.
3. You can define sections with 2D sketches (planar), 3D sketches (nonplanar), and planar or nonplanar faces, edges on a part, or points.
4. All sections must be either open or closed. You cannot mix open and closed profile types within the same loft operation. Open profiles result in a lofted surface.

Points

A point can be used as a section to help define the loft. The following rules apply to points used for a loft profile:

1. A point can be used to define the start or end of the loft.
2. An origin point, sketch point, center point, edge point, or work point can be used.

Rails

You can define rails by the following elements: 2D sketches (planar), 3D sketches (nonplanar), or part faces and edges. The following rules apply to rails:

1. There is no limit to the number of rails that you can create or include in the loft feature.
2. Rails must not cross each other and must not cross mapping curves.
3. Rails affect all of the sections not just faces or sections that they intersect. Section vertices without defined rails are influenced by neighboring rails.
4. All rail curves must be open or closed.
5. Closed rail curves define a closed loft, meaning that the first section is also the last section.
6. No two rails can have identical guide points, even though the curves themselves may be different.
7. Rails can extend beyond the first and last sections. Any part of a rail that comes before the first section or after the last is ignored.
8. You can apply a 2D or 3D sketch tangency or smooth constraint between the rail and the existing geometry on the model.

Centerline

A centerline is treated like a rail, and the loft sections are held normal to the centerline. When a centerline is used, it acts like a path used in the sweep feature, and it maintains a consistent transition between sections. The same rules apply to centerlines as to rails except that the centerline does not need to intersect sections and only one centerline can be used.

Area Loft

Select a sketch to be used as the centerline, and then click on the centerline to define the area of the profile at the selected point. After picking a point, the Section Dimensions dialog box appears, as shown in the following image. You define its position either as a proportional or absolute distance, and you can define the section's size by area or scale factor related to the area of the original profile.

FIGURE 7.67

Output. Select whether the loft feature will be a solid or a surface. Open sketch profiles selected as loft sections will define a lofted surface.

Operation. Select an operations button to add or remove material from the part, using the Join or Cut options, or to keep what is common between the existing part and the completed loft using the Intersect option. By default, the Join operation is selected.

Closed Loop. Click to join the first and last sections of the loft feature to create a closed loop.

Merge Tangent Faces. Click to join tangent faces on the completed loft into a single face.

Conditions Tab

The Conditions tab, as shown in the following image, allows you to control the tangency condition, boundary angle, and weight condition of the loft feature. These settings affect how the faces on the loft feature relate to geometry at the start and end profiles of the loft. This may be existing part geometry or the plane or work plane containing the loft section sketch.

FIGURE 7.68

Conditions. The column on the left lists the sketches and points specified for the sections. To change a sketch's condition, click on its name, and then select a condition option.

Condition Boundaries Sketches

Two boundary conditions are available when the first or last section is a point, as shown in the previous image.

- Free Condition (top button): With this option, there is no boundary condition, and the loft will blend between the sections in the most direct fashion.

- Direction Condition (bottom button): This option is only available when the curve is a 2D sketch. When selected, you specify the angle at which the loft will intersect or transition from the section.

Condition Boundaries Face Loop. When a face loop or edges from a part are used to form a section, as shown in the following image on the left, three conditions are available, as shown in the following image on the right.

FIGURE 7.69

- Free Condition (top button): With this option, there is no boundary condition, and the loft will blend between the sections in the most direct fashion.
- Tangent Condition (middle button): With this option, the loft will be tangent to the adjacent section of face.
- Smooth (G2) Condition (bottom button): With this option, the loft will have curvature continuity to the adjacent section of face.

Condition Boundaries Point. Three conditions are available when a point is selected for the loft profile, as shown in the following image.

FIGURE 7.70

- Sharp Point (top button): With this option, there is no boundary condition, and the loft will blend from the previous section to the point in the most direct fashion.
- Tangent (middle button): When selected, the loft transitions to a rounded or domed shape at the point.

- **Tangent to Plane (bottom button):** When selected, the loft transitions to a rounded dome shape. You select a planar face or work plane to be tangent to. This option is not available when a centerline loft is used.

Angle. This option is enabled for a section only when the boundary condition is Tangent or Direction. The default is set to 90° and is measured relative to the profile plane. The option sets the value for an angle formed between the plane that the profile is on and the direction to the next cross-section of the loft feature. Valid entries range from 0.0000001° to 179.99999°.

Weight. The default is set to 0. The weight value controls how much the angle influences the tangency of the loft shape to the normal of the starting and ending profile. A small value will create an abrupt transition, and a large value creates a more gradual transition. High weight values could result in twisting the loft and may cause a self-intersecting shape.

Transition Tab

The Transition tab, as shown in the following image with the Automatic mapping box unchecked, allows you to specify point sets. A point set is used to define section point relationships and control how segments blend from one section to the segments of the adjacent sections. Points are reoriented or added on two adjacent sections.

Map Points

You can map points to help define how the sections will blend into each other. The following rules apply to mapping points:

Point Set. The name of the point set appears here.

Map Point. The corresponding sketch for the selected point set appears here.

Position. The location of the selected map point appears here. You can modify the position by entering a new value or by dragging the point to a new location in the graphics window.

To modify the default point sets or to add a point set, follow these steps:

1. Click on the Transition tab, and uncheck the Automatic Mapping box. The dialog box will populate the automatic point data for each section. The list is sorted in the order in which the sections were specified on the Curve tab as shown in the following image on the left.
2. To modify a point's position, select the point set in which the point is specified. When you select its name, it will become highlighted in the graphics window.
3. Click in the Position section of the map point that you want to modify, and enter a new value or drag the point to a new location in the graphics window.
4. To add a point set, select Click to add in the point set area.
5. Click a point on the profile of two adjacent sections. As you move the cursor over a valid region of the active section, a green point appears. As the points are placed, they are previewed in the graphics window, as shown in the following image on the right.
6. You can modify the new point set in a similar way to the default point set.

Chapter 7 • Advanced Modeling Techniques   407

FIGURE 7.71

EXERCISE 7-5: CREATING LOFT FEATURES

In this exercise, you use the loft command to define the shape of a razor handle.

1. Open *ESS_E07_05.ipt* in the Chapter 07 folder.
2. In the browser, double-click Sketch1 to edit the sketch.
3. Click the Point, Center Point command on the Draw panel.
4. Click a point coincident with the spline near the bottom of the curve, making sure that the coincident glyph is displayed as you place the point, as shown in the following image on the left.
5. Draw two construction lines coincident with the sketched point and the nearest spline points on both sides of the Point that you just created, as shown in the following image on the right.

FIGURE 7.72

6. To parametrically position the sketched point midway between the spline points, place an equal constraint between the two construction lines.
7. Change the 4.750 overall horizontal dimension to **5 in**, and verify that the sketched point moves along the spline to maintain its position on the spline.
8. Click Finish Sketch on the Exit panel to finish editing the sketch.
9. Click the Work Plane command on the Work Features panel. To create a work plane perpendicular to the spline, click the Center Point you just created, and then click the spline, not the construction lines as shown in the following image on the left.
10. Next create a profile for a loft section on the new work plane. Create a new sketch on the new work plane.
11. Change to the Home View.

12. Project the Point, Center Point that you created in an earlier step onto the sketch.
13. Draw an ellipse with its center coincident with the projected point and the second point so the ellipse's axis is horizontally constrained to the Point. Click a third point, as shown in the following image.
14. Place **.1875 in** and **.375 in** dimensions to control the ellipse size, as shown in the following image on the right.

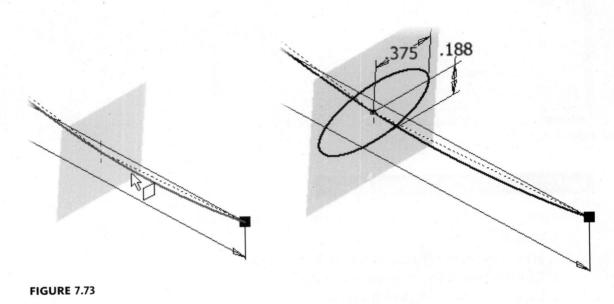

FIGURE 7.73

15. Finish the sketch.
16. Click the Loft command on the Create panel.
17. For the first section, click the concave 3D face, as shown in the following image.

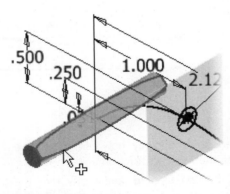

FIGURE 7.74

18. Click the other three profile sections from left to right, and then click the point on the right end of the spline.
19. Click the Center Line option in the dialog box, and then select the spline. When finished, the preview looks like the following image on the left.

FIGURE 7.75

20. Click OK to create the loft.
21. Turn off the visibility of all the work planes by pressing the ALT and] keys, and then use the Orbit command to examine the loft. Notice that the end is sharp at the point.

FIGURE 7.76

22. To edit the loft move the cursor over the Loft in the graphics window, and click the Edit Loft button from the mini-toolbar that is displayed. Click the Conditions tab, and change the Point entry to a Tangent condition, as shown in the following image on the left.
23. Click OK to update the loft. The right side of the part should resemble the following image on the right.

FIGURE 7.77

24. Next add another section that defines the section's area. In the browser, move the cursor over the entry Loft1 and double-click.

    - From the Curves tab, click Area Loft option.
    - In the Placed Sections area, click Click to add.
    - In the graphics window, click a point near the middle of the spline, as shown in the following image.

FIGURE 7.78

25. In the Section Dimensions dialog box:

    - Change the Section Position proportional distance to **0.5**.
    - Change the Section Size area to **0.375 in^2.**
    - Click OK.
    - In the graphics window, notice the values of each section.

26. In the Loft dialog box, click OK. The following image on the right shows the updated loft.

FIGURE 7.79

27. Close the file. Do not save changes. End of exercise.

MIRROR FEATURES

When creating a part that has features that are mirror images of one another, you can use the Mirror Feature command to mirror a feature(s) about a planar face or work plane instead of recreating the features from scratch. Before mirroring a feature, a work plane or planar face that will be used as the mirror plane must exist. The feature(s) will be mirrored about the existing plane; it can be a planar face, a work plane on the part, or an origin plane. The mirrored feature(s) will be dependent on the parent feature—if the parent feature(s) change, the resulting mirror feature will also update to reflect the change. To mirror a feature or features, use the Mirror command on the Model tab > Pattern panel, as shown in the following image on the left. The Mirror dialog box will appear, as shown on the right. The following sections explain the options in the Mirror dialog box.

FIGURE 7.80

Mirror Individual Features. Click this option to mirror a feature or features.

Mirror the Entire Solid. Click this option to mirror the solid body.

Mirror Plane. Click and choose a planar face or work plane on which to mirror the feature(s).

Solid. If there are multiple solid bodies, click this button to choose the solid body(ies) to receive the mirrored feature.

To mirror a feature or features, follow these steps:

1. Click the Mirror Feature command on the Pattern panel. The Mirror dialog box will appear.
2. Click the Mirror individual features option or the Mirror the entire solid option.
3. Select the feature(s) or solid body to mirror.
4. Click the Mirror Plane button, and then select the plane on which the feature will be mirrored.
5. Click the OK button to complete the operation.

The following image on the left shows a part with the features that will be re-created, the middle image shows the part with a work plane that the features will be mirrored about, and the image on the right shows the features mirrored.

FIGURE 7.81

SUPPRESSING FEATURES

You can suppress a model's feature or features to temporarily turn off their display, as shown in the following image on the left. Feature suppression can be used to simplify parts, which may increase system performance. This capability also shows the part in different states throughout the manufacturing process, and it can be used to access faces and edges that you would not otherwise be able to access. If you need to dimension to a theoretical intersection of an edge that was filleted, for example, you could suppress the fillet and add the dimension and then unsuppress the fillet feature. If the feature you suppress is a parent feature for other dependent features, the child features will also be suppressed. Features that are suppressed appear gray in the browser and have a line drawn through them, as shown in the following image on the right. A suppressed feature will remain suppressed until it is unsuppressed, which will also return the features browser display to its normal state. The following image on the right also shows the suppressed child features that are dependent on the suppressed extrusions.

FIGURE 7.82

To suppress a feature on a part, use one of these methods:

1. Right-click on the feature in the browser, and select Suppress Features from the menu, as shown in the previous image on the left.
2. Click the Select Feature option from the Quick Access toolbar. This option allows you to select features on the parts in the graphics window.
3. Right-click on the feature in the graphics window, and click the Suppress Features option from the menu.

To unsuppress a feature, right-click on the suppressed feature's name in the browser, and select Unsuppress Features from the menu.

REORDERING A FEATURE

You can reorder a feature in the browser. If you created a fillet feature using the All Fillets or All Rounds option and then created an additional extruded feature, such as a boss, you can move the fillet feature below the boss and include the edges of the new feature in the selected edges of the fillet. To reorder features, follow these steps:

1. Click the feature's name or icon in the browser. Hold down the left mouse button and click and drag the feature to the desired location in the browser. A horizontal line will appear in the browser to show you the feature's relative location while reordering the feature. The following image on the left shows a hole feature being reordered in the browser.

2. Release the mouse button and the model features will be recalculated in their new sequence. The following image on the right shows the browser and the reordered hole features.

FIGURE 7.83

If you cannot move the feature due to parent-child relationships with other features, Autodesk Inventor will not allow you to drag the feature to the new position. In the browser, the cursor will change to a No symbol instead of a horizontal line, as shown in the following image.

FIGURE 7.84

FEATURE ROLLBACK

While designing, you may not always place features in the order that your design later needs. Earlier in this chapter, you learned how to reorder features, but reordering features will not always allow you to create the desired results. To solve this problem, Autodesk Inventor allows you to roll back the design to an earlier state and then place the additional new features. To roll back a design, drag the End of Part marker in the browser to the location where the new feature will be placed. To move the End of Part marker, follow these steps:

1. Move the cursor over the End of Part marker in the browser.
2. With the left mouse button depressed, drag the End of Part marker to the new location in the browser. While dragging the marker, a line will appear, as shown in the following image on the left.
3. Release the mouse button, and the features below the End of Part marker are removed temporarily from calculation of the part. The End of Part marker will be moved to its new location in the browser. The following image in the middle shows the End of Part marker in its new location.
4. Create new features as needed. The new features will appear in the browser above the End of Part marker.
5. To return the part to its original state, including the new features, drag the End of Part marker below the last feature in the browser.
6. If needed, you can delete all features below the End of Part marker by right-clicking on the End of Part marker. Then click DELETE all features below EOP, as shown in the following image on the right.

FIGURE 7.85

Derived Part Symbols

The symbols show you how the solid bodies will be handled, the status of geometry that will be contained in the derived part, and whether or not it will be included or excluded from the derived part. The Derived Part Symbols table shows the available symbols.

DERIVED PART SYMBOLS

Symbol	Function	Meaning
	Merge Seams	Single solid body merging out seams between planar faces.
	Retain Seams	Creates a single solid body part that retains the seams between planar faces.
	Solid Bodies	If the source contains a single body, it creates a single body part. If the source contains multiple visible solid bodies, select the required bodies to create a multi-body part. This is the default option.
	Create Surface	Creates a part with the selected bodies as base surfaces.

Symbol	Function	Meaning
⊕	Include	Indicates that the selected geometry will be included in the derived part.
⊘	Exclude	Indicates that the selected geometry will not be included in the derived part. If a change is made to this type of geometry in the parent part, it will not be incorporated or updated in the derived part.
◐	Mixed	Indicates that the folder contains mixed included and excluded objects.

CONTENT CENTER

Autodesk Inventor's Content Center contains a number of standard components. It contains thousands of parts such as screws, nuts, bolts, washers, pins, and so on. You can place these standard components into an existing assembly using the Place from Content Center command that is found on the Assemble tab > Component panel, as shown in the following image on the left.

After the command is selected, the Place from Content Center dialog box opens, as shown in the following image on the right, and you can navigate between parts and features that are either included or published to the Content Center. Select the item you want to place, click OK, and place the part as you would any other assembly component.

FIGURE 7.86

In addition to using the default items in the Content Center, you can publish your own features and parts to the Content Center. You publish parts and features using the Editor or Batch Publish commands, found on the Manage tab > Content Center panel as shown in the following image.

FIGURE 7.87

In order to publish content to the Content Center, a read/write library must exist and be added to the active project file.

NOTE For more information about the Content Center, publishing features, publishing parts, or configuring libraries, refer to Autodesk Inventor's Help system.

DESIGN ACCELERATOR

Autodesk Inventor's Design Accelerator commands enable you to quickly create complex parts and features based on engineering data such as ratio, torque, power, and material properties. The Design Accelerator consists of component generators, mechanical calculators, and the *Engineer's Handbook*.

While working in an assembly the Design Accelerator commands are accessed on the Design tab. The commands found under the tab are as shown in the following image.

FIGURE 7.88

- Component and feature generators consist of bolted connections, shafts, gears, bearings, springs, belts, and pins.
- Mechanical calculators use standard mathematical formulas and theories to help you design components. The calculators available include weld, solder, hub joints, fit and tolerance, power screws, beams, and brakes.
- The *Handbook* contains engineering theory, formulas, and algorithms used in machine design.

To create parts and features with the Design Accelerator, follow these steps:

1. Open the assembly in which to place the part.
2. Activate the Design tab.
3. From the Design tab, select the operation you need to perform.
4. A wizard will walk you through the steps of the operation. An example of the Spur Gear Component Generator is shown in the following image.

FIGURE 7.89

> For more information about the Design Accelerator, refer to Autodesk Inventor's Help system.

NOTE

STRESS ANALYSIS

Autodesk Inventor Professional or Autodesk Inventor Simulation has the ability to determine the stress on a part, or how the part will react when certain forces are applied. Before a simulation can be performed, a material must be assigned, constraints established, and a load applied as shown in the following image on the left. Autodesk Inventor uses Finite Element Analysis (FEA) to create a grid across the surface called a mesh as shown in the following image in the middle. The mesh divides the object into a continuous set of finite elements. Using the properties of the assigned materials, Autodesk Inventor calculates the stress in each element as shown in the following image on the right.

FIGURE 7.90

Create Simulation

The first step when performing a stress analysis is to create a simulation. Inventor allows you to create more than one simulation on the same part or assembly. To create a simulation, click the Create Simulation command on the Environments tab > Manage panel, as shown in the following image on the left. The Create New Simulation dialog will appear, as shown in the following image on the right.

FIGURE 7.91

Before a simulation can be created, you select the type of simulation that will be created. Inventor allows a Static or Model Analysis to be simulated. The following descriptions explain the two types of simulation.

Static Analysis

This analysis performs a simulation that calculates stresses and displacement. It determines the simple structural loading conditions of the part without motion and the load is constant (no impact or changing of the load values).

Model Analysis

A Model Analysis is performed with the frequencies of vibration in consideration. Rigid body movements are in consideration within this type of analysis.

Assign Materials

A material can be assigned to a part in two ways. One way is to use the iProperties command on the Inventor Application menu, as shown in the following image on the left or right-click on the file's name in the browser and click iProperties from the menu. In the iProperties dialog click the Physical tab, and under the Material drop down list select the desired material as shown in the following image.

Chapter 7 • Advanced Modeling Techniques

FIGURE 7.92

The second method to assign a material is to use the Assign command on the Stress Analysis tab > Material panel, as shown in the following image on the left. The Assign Materials dialog will appear, as shown in the following image on the right. Under the Override Material drop down menu select the desired material.

FIGURE 7.93

Applying Constraints

After the material properties are assigned, a well-constrained solid model is needed before any analysis can be performed. These constraints represent the conditions the part or assembly will experience in the real world. Inventor has three basic constraints within the Stress Analysis environment. These constraints can be found under the Stress Analysis tab > Constraints panel, as shown in the following image.

FIGURE 7.94

Fixed, Pin, and Frictionless constraints constrain the part or assembly to assure that the model reacts to an applied force in a realistic manner. Before the proper constraint can be applied, you must determine how the constraints will limit the movement of the model. You should have a good understanding of the three constraints before deciding which to create a constraint. A brief description follows.

Fixed

A fixed constraint removes all the degrees of freedom on the selected face(s) or edge(s). You apply a fix constraint when no translational or rotational movements are permitted. You may apply multiple fix constraints on different faces or edges on the same model.

FIGURE 7.95

Pin

A pin constraint removes all freedoms along any axis of translation on a cylindrical face. No linear or sliding movement is allowed. Apply this constraint if rotational movement is acceptable. A pin constraint can only be applied to a cylindrical face. A Pin constraint can work in conjunction with the other constraints.

FIGURE 7.96

Frictionless

A Frictionless constraint limits a face to a rotation or translation movement along a plane. The face is prevented from moving or deforming perpendicular to the plane. A Frictionless constraint can work in conjunction with the other constraints.

FIGURE 7.97

Applying Loads

The final step in setting up a stress analysis is to apply a load to the model. Inventor has several different loads that can be applied to the model. These constraints can be found under the Stress Analysis tab > Loads panel, as shown in the following image. Review the following descriptions to determine the type of load that closely resembles the desired load that you need to apply.

FIGURE 7.98

Force

Applies a user-defined load to a face or edge. A Force load is perpendicular to the surface, and parallel to an edge. If a part that uses inches is simulated, the default unit of measure is lbforce (pound force). If a metric file is simulated, newtons will be used.

FIGURE 7.99

Pressure

Applies a user-defined pressure to a face. Pressure is a uniform load applied perpendicular to the whole surface. The default unit of measure is psi or MPa.

FIGURE 7.100

Bearing

Applies a user-defined load to a cylindrical face. A bearing load is parallel to the axis of the cylindrical face. The default unit of measure is lbforce or newtons.

FIGURE 7.101

Moment

Applies a user-defined rotational load around an axis. If applied to a face, the momentary load is perpendicular to the surface. The default unit of measure is lbforce in or newtons in.

FIGURE 7.102

Gravity

Applies a user-defined gravitational pull on a part. The pull is perpendicular to the surface, or parallel to an edge. The default unit of measure is in/s² or mm/s².

FIGURE 7.103

Simulate

After a material is assigned, loads and constraints are applied, a stress analysis can be simulated. Click the Simulate command from the Stress Analysis tab > Solve panel, as shown in the following image on the left. The Simulate dialog will appear, then in the Simulate dialog box click the Run button as shown in the flowing image on the right.

FIGURE 7.104

After a simulation is performed several results can be displayed. To view a different test, check the desired test result in the browser under the results section as seen in the following image.

FIGURE 7.105

Display

The results of the simulation are read by comparing the colors on the part to the color bar on left side of the Graphics Window. How the simulation is displayed can be manipulated by selecting one of the options under the Stress Analysis tab > Display panel, as shown in the following image.

FIGURE 7.106

- *Color Bar Settings:* Opens the Color Bar Settings Menu
- *Minimum Value:* Displays the location of the minimum value.
- *Maximum Value:* Displays the location of the maximum value.
- *Shading Display:* Allows user to change the quality of the shading.
- *Displacement Display:* Adjusts the level of model deformation.
- *Boundary Conditions:* Toggles the display of glyphs for the loads applied.
- *Probe Display:* Toggles the display of user probes.

Animate or Probe Results

Autodesk Inventor allows you to analyze the results through an animation or probe a specific point(s). These commands are located under the Stress Analysis tab > Results panel, as shown in the following image.

FIGURE 7.107

Animate

The Animate command plays an animation of the results. Once the Animate Results dialog opens, click the play button to view animation. If desired, an animation can be recorded.

Probe

The Probe command creates a probe to measure the simulation results. The probe will measure the exact value of the simulation at the selected point. To place a probe, select the area of the model where the exact measurement is to be displayed.

FIGURE 7.108

Report

Once the Simulation is complete, a report can be generated and exported. The Report command is located under the Stress Analysis tab > Report panel, as shown in the following image on the left. To determine what simulation results to include in the report, check the options desired on the simulations tab in the report dialog as shown in the following image on the right. From the Format tab, select the file type to export; html, mhtml, or rtf. Click the OK button to create the report.

FIGURE 7.109

EXERCISE 7-6: RUNNING A STRESS ANALYSIS ON A PART

In this exercise, you will use the Stress Analysis commands to create a simulation, assign material, constraints, loads, simulate, animate, and create a report of a part.

1. Open *ESS_E07_06.ipt* in the Chapter 07 folder.
2. Click the Environments tab.
3. From the Begin panel, click the Stress Analysis command as shown in the following image on the left.
4. Click the Create Simulation command, as shown in the following image on the right.

FIGURE 7.110

5. In the Create New Simulation dialog box that appears, click OK.
6. A material has not been applied to this part via the iProperties. Assign a material to the part for the simulation by clicking Stress Analysis tab > Material panel > Assign and assign a material in the override material drop down list click Aluminum-6061, as shown in the following image and then click OK.

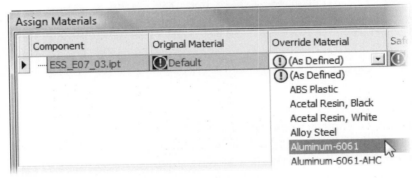

FIGURE 7.111

7. Next you constrain the part, click the Fixed command from the Stress Analysis tab > Constraints panel.
8. To create a fixed constraint on the front end of the part, select the front circular face as shown in the following image and then click OK.

FIGURE 7.112

9. Next you apply a load to the part, click the Force command from the Stress Analysis tab > Loads panel.
10. Select the opposite circular face of the part, change the Magnitude to **50 lbforce**. Flip the direction of the force, click the Direction button so the force is directed away from the part as shown in the following image. Click OK to create the force.

FIGURE 7.113

11. Next, you run the simulation. Click the Simulate command from the Stress Analysis tab > Solve panel and then click the Run button in the Simulate dialog box.
12. The result of your simulation should resemble the following image.

FIGURE 7.114

13. From the browser, review the results of the simulation.
14. Next, you animate the simulation results. Click the Animate command from the Stress Analysis tab > Result panel. To start the animation, click the Play button in the Animate Results dialog box as shown in the following image. You can adjust the speed by selecting different options from the drop list. Note that Animations can be recorded by clicking the Record button in the Animate Results dialog box.

FIGURE 7.115

15. Click OK to finish the animation.
16. Lastly, you generate a report. Click the Report command from the Stress Analysis tab > Report panel and then click OK.
17. To return to the part file environment, click the Finish Stress Analysis command from the Stress Analysis tab > Exit panel.
18. Close the file. Do not save changes. End of exercise.

EXERCISE 7-7: RUNNING A STRESS ANALYSIS ON AN ASSEMBLY

In this exercise, you will use the Stress Analysis commands to create simulation, assign material, constraints, loads, simulate, animate, and create a report on an assembly.

1. Open *ESS_E07_07.iam* in the Chapter 07 folder.
2. Click Stress Analysis command from Environments tab > Begin panel.
3. Click the Create Simulation command from the Stress Analysis tab > Manage panel. In the Create New Simulation dialog box, click the OK button.
4. Next you apply a material to each part. Click the Assign command from the Stress Analysis tab > Manage panel and assign the material Aluminum 6061 to both parts. You only need to change one occurrence of each part in the Assign Materials dialog box.
5. Next, you fix the assembly. Click the Fix command from the Stress Analysis tab > Constraints panel. Fix the four bottom circular faces (not the edges) of the assembly as shown in the following image.

FIGURE 7.116

6. Next, you apply a pressure. Click the Pressure command from the Stress Analysis tab > Loads panel. Apply a **20.000 psi** pressure to the top-two planar faces of the parts as shown in the following image.

FIGURE 7.117

7. Run the simulation by clicking the Simulate command from the Stress Analysis tab > Solve panel and click Run in the Simulate dialog box.
8. From the browser, review the results of the simulation.
9. If desired, animate the results and generate a report.
10. To return to the part file environment, click the Finish Stress Analysis command from the Stress Analysis tab > Exit panel.
11. Close the file. Do not save changes. End of exercise.

APPLYING YOUR SKILLS

Skill Exercise 7-1

In this exercise, you use the knowledge you gained through this course to create a joystick handle. You will use the loft, split, and a few plastic part commands.

1. Open *ESS_Skills_7-1.ipt* from the Chapter 07 folder.
2. Create a loft, for the sections use the three elliptical shapes in order from top to bottom and for the rails use the splines and your screen should resemble the following image on the left.
3. Create a **.1875 in** fillet around the top edge, as shown in the following image on the right.

FIGURE 7.118

4. Shell the part to the inside with a thickness of **.0625 in**. Do not remove any faces.
5. Close the file. Do not save changes. End of exercise.

Skill Exercise 7-2

In this exercise, you use the knowledge you gained through this course to perform a stress analysis on a part. You will use the Stress Analysis environment and the create simulation, assign material, constraints, loads, simulate, animate, and report commands.

1. Open *ESS_Skills_7-2.ipt* from the Chapter 07 folder.
2. Create a Stress Analysis.
3. Assign Alloy Steel as the material to the part.
4. Apply the Fixed constraint to the top-rectangular face of the part as shown in the following image.

FIGURE 7.119

5. Apply a load using the Moment command to apply a moment to the front face with a value of **50.000 lbforce in,** as shown in the following image.

FIGURE 7.120

6. Run the stress analysis simulation.
7. From the browser, review the results of the simulation.
8. Animate the results and generate a report.
9. While in the Stress Analysis environment, from the browser, right-click on the Loads > Moment:1 entry and click Edit Moment and enter **100.000 lbforce in** and rerun the simulation.
10. Click Finish Stress Analysis to return to the part environment.
11. Close the file. Do not save changes. End of exercise.

CHECKING YOUR SKILLS

Use these questions to test your knowledge of the material covered in this chapter.

1. True ___ False ___ You can only emboss text on a planar face.
2. True ___ False ___ A sweep feature requires three unconsumed sketches.
3. True ___ False ___ In a 3D Sketch the Project Geometry command is used to project curves onto a circular face.
4. True ___ False ___ You can create a 3D curve with a combination of both 2D and 3D curves.
5. Explain how to create a 3D path using geometry that intersects with a part.
6. True ___ False ___ You can control the twisting of profiles in a loft by defining point sets.
7. Explain the difference between suppressing and deleting a feature.
8. True ___ False ___ After mirroring a feature, the mirrored feature is always independent from the parent feature. If the parent feature changes, the mirrored feature will NOT reflect this change.
9. Explain how to reorder features.
10. True ___ False ___ When placing text on an arc with the Geometry Text command the text will start at the left quadrant of the arc.
11. When creating a sweep with the Path & Guide Surface option, what does the surface control?
12. True ___ False ___ The Hardware Center contains standard parts, such as nuts and bolts, that can be inserted into an assembly.
13. True ___ False ___ A stress analysis simulation can be run without assigning a material to the part(s).
14. True ___ False ___ A fixed constraint must be applied to the part or the assembly before a stress analysis can be run.
15. What does the color bar indicate after a stress analysis simulation has been run?

INDEX

A

adaptivity
    exercise, 304–307
    main discussion, 301–302
    when driving constraints, 321
aligned sections, 207–208
alignment, breaking, 221
ALT + drag constraining, 284
Alternate Solution option, 98, 106
analysis, stress. *See* stress analysis
analysis commands, 315–317
angle constraints, 279
angles
    dimensioning, 78
    loft, 406
    measuring, 60
animating
    constraints, 319–324
    DOF, 273–274
    in presentations, 325, 329–330
    simulation results, 424–425
annotations. *See also* balloons; BOM (Bill of Materials); dimensions; parts lists; text
    additional commands, 254–255
    center marks and centerlines, 227–231. *See also* centerlines
    exercise, 257–260
    hole and thread notes, 255–256
appearance options, 35–37
application options, 13–14. *See also specific options*
arcs
    dimensioning, 78–79
    exercise, 64–65
    sketching, 54–55
Area Loft option, 403
areas, measuring, 60
assemblies. *See also* assembly constraints; BOM (Bill of Materials); components; presentation (.ipn) files; stress analysis
    analysis commands, 315–317
    capacity meter, 270
    creating, 265
    drawing views from, 336
    exercises
        analysis, 317–319
        assembling parts, 288–295
        designing parts in assembly context, 304–307
    files (.iam), 9, 265
    option settings, 266
    section tools, 303
Assembly browser
    commands, 299–301
    overview, 269–270
    views, 296
assembly constraints
    additional commands for, 295–299
    additional ways to apply, 283–284
    driving, 319–321
    editing, 286–287
    exercise, driving, 322–324
    isolating errors, 298–299
    limits on, 276–277
    offset value modification, 287
    options for, 275–276
    sets, 282–283
    terminology, 274
    transitional, 281–282
    types of, 274–275, 277–281
Assembly View, 296
assigning materials, 418–419
associated patterns, 308–309
auto ballooning, 338–341
Autodesk Inventor Publisher, 330–331
Autodesk Vault, 21, 24
automated centerlines, 230–231
automatic constraints, 56
autoprojecting, 45, 49, 119–120
auxiliary views, 194, 204–205, 213–216
AVI rate, 321
axes
    visibility of, 48–49
    work, 157–161

B

balloons
    discussion, 336–343
    exercise, 348–353
base features, 87–88
baseline dimensions and sets, 246–248, 251–253
base views, 193, 194–198
bearing loads, 422
BOM (Bill of Materials)
    editing from parts lists, 344–345
    settings for balloons, 338, 339
bottom-up assembly approach, 265
boundary conditions, 404–405
break/broken views, 194, 216–219
break out views, discussions, 194
browser, 89–90. *See also* Assembly browser

C

Cartesian coordinates, 52–53
centered patterns, 229–230
centerline bisectors, 228–229
centerlines
    for drawings, 227–234
    for loft features, 403
    for revolving sketches, 100–101
center marks, 229
center of gravity, 316
Center-Point Arc/Circle commands, 57
center points, 48–49, 147–148
chain dimensions and sets, 249–253
chamfers
    command overview, 58
    exercise, 140–143
    main discussion, 137–140
circles, dimensioning, 78–79
circular patterns
    creating, 179–181, 310
    exercise, 183–184
    overview, 58, 175
clearance holes, 147
closed profiles, embossing, 363–372
coincident constraints, 66
collinear constraints, 67
collision detection, 321
colors
    of Assembly tab selections, 275
    of parts, 37, 124
commands. *See also specific commands and objects*
    on Quick Access Toolbar, 6
    working with, 17–20
components
    copying, 269–270
    creating, 267–269
    editing, 270–272
    exercises
        patterning, 312–315
        tweaking, 332–335
    grounded, 272
    isolating, 297–298
    moving and rotating, 285–286
    patterning, 308–312
    placing, 267
    tweaking, 326–329
Component tab, drawing views, 195–196
Compute options for rectangular patterns, 178
concentric constraints, 66
Concentric placement, 145
Conditions tab, loft feature, 404–406
Connection Line option, 212
constraints. *See also* assembly constraints; dimensions
    adding, 67
    automatic, 56
    deleting, 69
    exercise, 72–76
    overview, 65–66
    scrubbing, 57
    for stress analysis, 419–421
    types of, 66–67
    visibility of, 68–69
constraint sets, 282–283
construction geometry, 69–72
consumed sketches, 88
Content Center, 415
coordinate types, changing, 52–53
copies, saving, 12
creating objects. *See under specific objects*
creation methods for circular patterns, 180
cropped views, 194
curves, projecting, 396–397
Curves tab, loft feature, 402–404
custom balloons, 342–343
custom parts, 347–348
Cut button. *See* Operation options
cut edges, projecting, 120

D

degrees of freedom, 71, 272–274
deleting objects, 59. *See also under specific objects*
derived part symbols, 414–415
Design Accelerator, 416–417
Design Doctor, 286, 298–299
design view representations, 362
detail views, 194, 210–212, 213–216
dialog boxes, maximizing/minimizing, 19–20
dimension input, 54
dimensions
    auto-arranging, 237
    baseline, 246–248

2D sketches. *See* sketches
3D paths/sketches
    discussion, 395–398
    exercise, sweeping, 398–401

433

Index

chain, 249–251
constraint, 286–287
display, 372–373
for drawing views, 189
driven, 81–82
editing, 79–80, 236
exercises
  adding, 241–246
  creating baseline and chain, 251–253
  dimensioning a sketch, 83–85
  relationships and parameters, 381–387
general, 76–79, 238–239
linear diameter, 100–102
moving and centering text, 238
overview, 76
persistent, 53
relationships between, 373–381
repositioning, 81
retrieving, 234–236
selecting, 236–237
direction condition, 405
Direction options
  for holes, 146
  for rectangular patterns, 177–178
  for shell features, 154
  when extruding sketches, 97
  when revolving sketches, 105
  when tweaking components, 327
direct manipulation tools, 91–92
direct model edge referencing, 119
display options, for drawing views, 197–198
distance
  for chamfers, 138–139
  for extents, 96
  obtaining minimum, 315
documenting, using Inventor Publisher, 330–331
documents. *See* files
DOF, 71, 272–274
drawings and drawing views
  annotations. *See* annotations
  auxiliary views, 204–205
  base views, 194–198
  breaking alignment, 221
  break views, 216–219
  creating drawings, 190
  creating drawing views, 336
  deleting, 220–221
  detail views, 210–212
  dimensions. *See* dimensions
  drawing sheet preparation, 191–192
  DWG TrueConnect, 190–191
  editing properties of, 219–220
  exercises
    auxiliary, section, and detail views, 213–216
    break and break out views, 218–219
    editing, 222–224
    multiview drawing, 199–203
  files (.dwg and .idw), 9
  moving, 212
  opening from models, 261
  opening models from, 260–261
  options for, 189–190
  overview, 188
  replacing model reference, 221–222
  saving as template, 193
  section views, 205–210
  standards, 225
  styles. *See* styles
  types overview, 193–194
drawing sheets, 191–192
drill point types, 146
driven dimensions, 81–82
driving constraints, 319–324
dual unit, 61
DWG files, 9. *See also* drawings and drawing views
DWG TrueConnect, 190–191
Dynamic Input, 52
dynamic rotation, 33–34, 35

E

Edge Chain option, 139–140
edge fillets, 132–134
edge orientation, 33
edges
  direct manipulation tools for, 92
  projecting, 45, 119–121
editing objects. *See under specific objects*
embossing text and closed shapes, 363–372
environments, switching, 90
equal constraints, 67
equations, dimensions, 374, 376
Excel spreadsheets, as linked parameters, 375, 378–381
exercises. *See under specific objects*
Explosion Method options, 326
Extend command, 58
Extents options
  when extruding sketches, 95–97
  when revolving sketches, 104–105
extruding sketches
  discussion, 93–98
  exercise, 99–100

F

face fillets, 135
faces
  cycling, 116
  direct manipulation tools for, 91
  projecting curves onto, 396–397
face viewpoint, 33
features. *See* loft features; placed features; sketched features; sweep features; work features
  deleting, 112
  direct manipulation tools for, 91–92
  editing, 109–114
  failed, 112–113
  mirroring, 411–412
  overview, 87–88
  renaming, 112
  reordering, 413
  rollback, 413–414
  suppressing, 412
files. *See also specific types*
  creating, 8
  opening, 7
  opening multiple, 10–11
  saving, 11–12
  types of, 9–10
fillets
  creating, 130–131
  edge, 132–134
  editing, 131–132
  exercise, 140–143
  face, 135
  full round, 136
  overview, 58, 130
  reordering, 413
fix constraints, 67
fixed constraints, 420
flat patterns, projecting, 120
force loads, 421
Format Text dialog box, 254
fractions, 80
free condition, 404
frictionless constraints, 420–421
From Sketch placement, 145
Full Detail Boundary option, 212
full round fillets, 136

G

general dimensions, 76–79, 238–239
geometric constraints. *See* constraints

geometry
  creating 3D paths from, 395–396
  creating text about, 364–365
  projecting, 119–123
gesture behavior, 18–19
gravity loads, 423
grounded components, 272
guide rails, 390
guide surfaces, 391

H

hatching, 197, 208–210
Heads-Up Display (HUD), 52, 53–54
help system, 20–21
hole notes, 255–256, 257–260
holes
  discussion, 143–148
  exercise, 148–152
home view, 29–30, 32, 34. *See also* isometric views
horizontal constraints, 67

I

.iam files, 9, 265. *See also* assemblies
.idw files, 9. *See also* drawings and drawing views
Infer iMates option, 98, 107, 146
inferred points, 56
in-place activation, 299–300
in-place assembly approach, overview, 265
insert constraints, 280
interference checking, 316–317
Intersect button. *See* Operation options
Inventor Application Menu, 4–5
.ipj files, 10. *See also* projects
.ipn files, 9. *See also* presentation (.ipn) files
iProperties dialog box, 124
.ipt files, 9, 265
isometric views
  adding dimensions to, 239–240
  hatching, 209–210
  projected, 193, 199
  switching to, 29–30, 32–33

J

Join button. *See* Operation options
justification, view, 198

K

keys, shortcut, 19

L

leader text, 253–254
library search paths, 26
lights, 36
linear diameter dimensions, 100–102
linear patterns, 181
Linear placement, 145
lines
  defined, 274
  dimensioning, 77
  exercises, 62–65
  for revolving sketches, 100–102
  sketching, 54–55
linked parameters, 375, 378–381
loads, 421–423
Lock to Current Selection option, 34

loft features
- discussion, 401–407
- exercise, 407–410

loops
- measuring, 60
- projecting, 121

M

map points, 406–407
marking menus, 18–19
Match Shape option, 97, 106
mate constraints, 277–279
Measure commands, 59–62
measurement units. *See* units
menus
- application, 4–5
- marking, 18–19

minimum distance, obtaining, 315
Minimum Solution option, 98, 106
mini-toolbars, 91–92
mirroring
- command overview, 58
- features, 411–412

model analysis, 418
Modeling View, 296
model parameters, 375, 376
model references, replacing, 221–222
models
- opening drawings from, 261
- opening from drawings, 260–261

Model State tab, 196–197
momentary loads, 422
More options
- when extruding sketches, 97–98
- when revolving sketches, 106–107

motion constraints, 280–281
Motion tab, 276
moving objects. *See under specific objects*

N

Navigation Bar, 30–32
New File dialog box, 8
New Solid button. *See* Operation options
nonlinear paths, patterning exercise, 184–185
normal, defined, 274
normal lines, for revolving sketches, 101–102
notes, hole and thread, 255–256. *See also* text

O

oblongs, creation exercise, 117–119
Offset command, 58
offsets, defined, 274
On Point placement, 146
Open dialog box, 7
Operation options
- for loft features, 404
- for sweep features, 389
- when extruding sketches, 94–95
- when revolving sketches, 104

options, application, 13–14. *See also specific options*
Orbit commands, 31
Orientation options
- for drawing views, 195
- for rectangular patterns, 178–179
- for sweep features, 389

origin 3D indicator, 49
orthographic views, 34, 36–37, 193, 199, 209
Other Half command, 296–297
overconstrained sketches, 67, 81–82
overlay views, 194

P

Pack and Go command, 12
Pan command, 31
panels, issuing commands from, 17–18
parallel constraints, 66
parameters
- discussion, 375–381
- exercise, 381–387

parent-child relationships, 89
parts and part modeling. *See also* components; features; sketches; stress analysis
- creating, 47–48
- design view representations, 362
- embossing/engraving, 363–366
- exercises
    - embossing/engraving, 367–372
    - shelling, 155–157
- files (.ipt), 9
- material, properties, color, 124–125
- options for, 44
- sectioning, 361–362

parts lists
- discussion, 343–348
- exercise, 348–353

paths
- patterns along, 181
- sweep, 388–391

patterns
- along a path, 181
- associated, 308–309
- circular. *See* circular patterns
- editing, 310–311
- exercises
    - along nonlinear path, 184–185
    - circular patterns, 183–184
    - components, 312–315
    - rectangular patterns, 181–182
- linear, 181
- overview, 175
- rectangular. *See* rectangular patterns
- replacing component in, 311–312

perpendicular constraints, 66
persistent dimensions, 53
perspective views, 34, 36–37
pin constraints, 420
placed features. *See also* chamfers; features; fillets; holes; work features
- overview, 129
- shelling, 153–157

placement
- for balloons, 339, 340–341
- for circular patterns, 180
- for holes, 144–146

planes. *See also* work planes
- for active sketch, 48–49, 115–116
- defined, 274
- visibility of, 48–49

Point Alignment On option, 56
pointer input, 54
points
- center, 48–49, 147–148
- defined, 274
- inferred, 56
- for loft features, 402–403, 406–407
- work, 161–162

Polar coordinates, 52–53
Polygon command, 58
positioning methods for circular patterns, 180–181
Precision option, 61
presentation (.ipn) files
- drawing views from, 336
- exercise, 332–335
- main discussion, 324–330
- overview, 9

Preserve All Features option, 134, 140
pressure loads, 421–422
previews, options for, 190, 196, 275
probes of simulation results, 424–425
profile scaling options, 390
projected views, 193, 199

projecting
- to 3D sketch, 398
- curves onto a face, 396–397
- edges, 119–121
- geometry, 119–123

projects
- creating, 22–26
- exercise, 27–29
- files (.ipj), 10
- file search options, 22
- overview, 21
- setup, 22

Q

Quick Access Toolbar, 6
Quick Launch, 9

R

Radius Threshold settings, 231
rails, loft feature, 403
ray tracing, 36
rectangular patterns
- creating, 176–179, 309
- exercise, 181–182
- overview, 58, 175

rectangular text, creating, 363–364
redoing commands, 20
reference dimensions, 81
reference parameters, 375, 376
reflections, 36
Region Properties, 61–62
reordering features, 413
repeating last command, 19
reports, simulation, 425
ribbon, 5–6, 17–18
rollback, feature, 413–414
Roll options for fillets, 134
rotating viewpoint, 33–34, 35
rotation and rotation-translation constraints, 280–281
rounds, 130. *See also* fillets

S

Save commands, 11–12
scale options, drawing view, 195–196
scaling profiles, 390
scrubbing, 57
sectioning assemblies, 303
sectioning part files, 361–362
sections, loft feature, 402
section views
- discussions, 194, 205–210
- exercise, 213–216

selecting objects, 58–59. *See also under specific objects*
Select Other command, 116
Setback option, 139–140
shadows, 36
Shape options
- when extruding sketches, 94–97
- when revolving sketches, 103–106

sheet metal design files (.ipt), 9
shelling, 153–157
shortcut keys, 19
simple holes, 147
simulation, 417–418, 423–425
single-user projects, creating, 24–26
sketched features. *See also* features; sketches
- creating
    - basic steps for, 114–115
    - by extruding a sketch, 93–98
    - by revolving a sketch, 100–107
- exercises
    - creating an oblong, 117–119
    - editing features and sketches, 113–114

extruding a sketch, 99–100
revolving a sketch, 107–109
overview, 87
sketches. *See also* 3D paths/sketches; sketched features
  active, 48, 50, 115–116
  constraining. *See* constraints
  consumed, 88
  creating, 50
  dimensioning. *See* dimensions
  direct manipulation tools for, 91
  dragging, 72
  editing feature's, 110–111
  exercises
    dimensioning, 83–85
    using lines, 62–64
    using tangencies, 64–65
  finishing, 90
  options for, 44–45
  overview and guidelines, 50–51
  plane selection, 48–49, 115
  projecting to 3D, 120
  tools and techniques for, 51–62
  unconsumed, 88
sketch planes, setting default. *See also* planes
Slice Graphics command, 117
smooth (G2) constraints, 66
Smooth Cutout Shape option, 212
snaps, 70–71
spreadsheets, as linked parameters, 375, 378–381
static analysis, 418
stress analysis
  animating or probing results, 424–425
  applying constraints, 419–421
  applying loads, 421–423
  assigning materials, 418–419
  creating simulation, 417–418
  exercises, 426–429
  overview, 417
  reports, 425
  simulating, 423–424
styles
  drawing, 225–227
  drawing view, 196
  visual, 35–36
subassemblies, overview, 265
suppressing features, 412
surface texture symbols, exercise, 257–260
sweep features
  3D sketching, 395–398
  creating, 391–392
  dialog box, 388–391
  exercises
    3D sketches, 398–401
    creating, 392–395
  overview, 388
symmetry constraints, 67

T

tabs, document, 10–11
Tangent Arc/Circle commands, 57
tangent constraints, 66, 279
tangents, dimensioning to, 78–79
Taper option, 98, 390
tapped and taper tapped holes, 147
templates, 12, 46–47, 190, 193
termination options for holes, 146
text
  adding to drawing, 253–254
  creating about geometry, 364–365
  creating rectangular, 363–364
  dimension, 238
  embossing/engraving, 363–366
  exercises
    adding, 257–260
    embossing/engraving, 367–372
thread notes, 255–256
Three-Point Arc command, 57
Three-Point Rectangle command, 58
title blocks, 192
toolbar, quick access, 6
tooltips
  configuring, 17–18
  for constraints, 297
transitional constraints, 281–282
Transitional tab, constraint, 276
Transition tab, loft feature, 406–407
Trim command, 58
tweaking components, 326–329
Two-Point Rectangle command, 57

U

UCSs
  constraining together, 276, 282–283
  exercise, 169–175
  main discussion, 166–168
ul (unitless), 374
unconsumed sketches, 88
under-constrained adaptive features, 301–302
undoing commands, 20
unique face thickness, 153, 155
unitless numbers, 374
units
  dual, 61
  in equations, 374
  overriding default, 46
  for parameters, 376
Universal Naming Convention (UNC) paths, 23

user coordinate systems. *See* UCSs
user interface. *See also specific concepts*
  default sketch environment for parts, 2–4
  exercise, 14–17
user parameters, 375, 376, 377–378

V

Vault projects, 24
vertical constraints, 67
ViewCube, 29–30, 32–35
View Face command, 31–32
viewpoints
  exercise, 37–42
  Navigation Bar, 30–32
  overview, 29
  rotating, 33–34, 35
  ViewCube, 29–30, 32–35
views, drawing. *See* drawings and drawing views
visibility
  of axes, 48–49
  of center point, 48–49
  of components in browser, 300–301
  of constraints, 68–69
  of DOF, 71, 273
  of planes, 48–49
  of work features, 168–169
visual styles, 35–36

W

weldments, 196, 197
work axes, 157–161, 309
work features. *See also* patterns; UCSs; work planes
  overview, 157
  work axes, 157–161, 309
  work points, 161–162
work planes
  discussion, 162–169
  exercise, 169–175
work points, 161–162

Z

Zoom commands, 31